Community Learning & Libraries
Cymuned Ddysgu a Llyfrgelloedd

Newport
CITY COUNCIL

This item should be returned
last date stamped below.

V   HE
        R

To renew visit:

www.newport.gov.uk/libraries

David Garner, *Last Punch of the Clock*, 2009. (*Amgueddfa Cymru* National Museum Wales – copyright belongs to the artist)

VOICES FROM THE FACTORY FLOOR

The Experiences of Women Who Worked in
the Manufacturing Industries in Wales,
1945–75

CATRIN STEVENS

AMBERLEY

To all the wonderful 'factory girls' who shared their fascinating stories with us

First published 2017

Amberley Publishing
The Hill, Stroud
Gloucestershire, GL5 4EP

www.amberley-books.com

British Library Cataloguing in Publication Data.
A catalogue record for this book is available from the British Library.

ISBN 978 1 4456 4972 6 (print)
ISBN 978 1 4456 4973 3 (ebook)

Typeset in 10pt on 13pt Sabon.
Origination by Amberley Publishing.
Printed in the UK.

Contents

Introduction

Over the years, women's history has been a Cinderella topic, neglected and ignored by most Welsh historians. Recently, however, the academic works of Deirdre Beddoe, Angela John, Ursula Masson and others have begun to address this issue[1], and pioneers such as feminist Margaret Haig Thomas, Crimean nurse Betsi Cadwaladr and industrialist Amy Dillwyn have been rescued from the shadows of history. Women's Archive of Wales *Archif Menywod Cymru* was established in 1997 to further this cause, to raise the profile of women's history in Wales, and to identify and preserve sources relating to this history. The project that informs the writing of this book arose from the deliberations of Women's Archive of Wales, and received funding from the Heritage Lottery Fund and donations from the Ashley Family Foundation, trade unions – Union and Community – and from our own members. The topic chosen was an examination of the lives of working-class women who worked in the manufacturing industries in Wales between 1945 and 1975. It aimed at capturing their memories and experiences before it was too late.

As the title *Voices from the Factory Floor* suggests, the research was conducted through recording the oral histories of over 200 speakers. Oral history has its champions and detractors, but no other source would have enabled us to tap so directly into such a wide range of ordinary stories. Their being female, working-class voices could have rendered them worthless in the eyes of many historians, and they could have remained hidden forever. This project has given us the incredible privilege of being able to listen to their many voices, and to appreciate their many histories. Although, individually, they are fragments of history, together they form a formidable narrative, which historians of this under-examined period of Welsh history cannot afford to ignore or discount.

Some interviewees were aware of historians' dilemma with oral history. Mair Griffiths (Voices North, VN033) who worked in Llangefni Creamery, Anglesey, admits that, '*You look back and it's all honey, but …*', while Sheila Edwards (VN049), Graesser's Salicylates, Sandycroft, wonders whether she is seeing her working life 'through rose coloured spectacles'. The recordings prove, however, that the speakers are not practised manipulators of facts and feelings; any inconsistencies, such as with dates or even the names of factories (which seemed to change ownership almost overnight without affecting the workforce), were genuine mistakes or oversight. A far bigger hurdle to overcome was the fact that many of the potential interviewees believed that they had little to contribute. They claimed their lives had been mundane and unremarkable. Yet, when they started to talk and when half an hour had flown past, they realised that they had a valid and important story to tell.

We discovered that they enjoyed the experience; that they could express their views and emotions lucidly, and with humour or poignancy as befitted the occasion. Sixty of the 205 interviewees were recorded in Welsh and they are quoted in the text in italic font.

The post-war period explored in this book was one of gradual decline for the traditional, male-orientated, heavy industries of coal, steel and slate in Wales. Women had been drawn or forced in huge numbers into the workplace during the Second World War but seemed destined to return to domesticity in 1945. Indeed, between July 1945 and July 1946, the number of unemployed women in Wales increased from 6,835 to 29,079[2]. The Distribution of Industry Act of 1945 sought to address the issue of the large reserves of female labour and, through the interventionist policies of its post-war reconstruction programme, the government aimed at creating 125,000 new jobs (70,000 for men and 55,000 for women)[3]. Thus, British companies were enticed to open branches on new industrial estates at Treforest, Hirwaun, Bridgend, Swansea and Wrexham. By 1955, 34,300 women (and 32,050 men) were employed in the numerous factories on these estates, with the numbers remaining fairly constant and reaching their peak in 1973 with 74,469 workers – 35,694 of whom were women[4]. Many other companies, some of them native, opened factories outside these main conglomerations. Male workers formed a significant portion of this workforce, and their stories also need to be recorded and appreciated; however, our remit and emphasis in this project was to focus upon the female cohort and rescue its history from oblivion.

As hundreds of factories opened, and several closed their doors in Wales during this period, it is extremely difficult to establish a true record of events. The interviewees had worked in over 200 factories between them and these varied in size. Large enterprises included, for example, British Nylon Spinners, Pontypool (4,000–5,000 workers with a fair proportion of women) or AB Metals, Abercynon, with *c.* 4,000 female employees, while medium-sized enterprises included the numerous Laura Ashley factories in mid Wales or Revlon in Maesteg with 400–500 employees; there were also small units such as Austin Taylor's Bethesda (*c.* 100 employees) and rural Deva Dog Ware, Gwynfe, Carmarthenshire, with merely ten to fifteen workers. More confusingly, the factories changed hands frequently. As Sonia Gould (Voices South East VSE002) explains, although the factory she worked in changed hands from Denex to Harvey's to Clifford Williams & Son, it was always known locally in Tredegar as 'Denex'.

Discussion on the size of the workforce proved to be one of the most taxing questions for the interviewees. Such vague estimates as 'several hundreds', 'thousands', 'loads' and even 'a tidy size' were offered, but this uncertainty probably reflects the fact that numbers went up and down almost at random and that they were familiar with only their own section of the factory. They had little doubt, however, that during this period there was plenty of work, as Vicky Perfect (VN028) who worked at Courtauld's, Flint, comments:

> In 1964, when I started work, you could get one job in the morning, another one dinner time if you didn't like that one, another one shortly after dinner, and a fourth one before the day was over. It was booming.

Mary Evans (VN008) did just that. She fell out with her supervisor at the Corset Factory in Caernarfon one morning in 1956 and so marched down to James Kaylor Compacts, where she was told to 'Start Monday'. According to Sylvia Reardon (VSE006), the *South*

Wales Echo would be 'crammed' with job advertisements – you could apply for four and be offered three – while Windsmoor's, Swansea, posted notices outside when it was 'taking on' new workers (Voices South West VSW052). Little wonder, therefore, that some of the interviewees had worked in a bewildering number of factories. Between 1962 and 1998, Patricia Ridd (VSW041) worked in six different factories on the Fforestfach Industrial Estate, Swansea, including Britvic Corona for a period of twenty-six years. Likewise, Patricia Howard (VSE029) moved between six factories between 1958 and 1996, with one brief sojourn in the Winchester Sausage Factory in England in the early 1960s. Their chequered histories sometimes confused the interviewees themselves as they struggled to sequence their working lives. Yet this did not diminish their capacity to describe their experiences in each factory and in no way did it invalidate their testimony. They were well aware that this was an exciting new era for light industry and that they were in the vanguard as industrial workers. Shirley Smith (VSE034) proudly reeled off the names of the factories she encountered as she walked to work at Burry, Son & Co. on the Treforest Estate in the 1960s:

> We'd walk through GECs, and then Oh! Liners' was there, and the smell from Liners' – Oh, the bone factory! ... DCL and then Chrome Leather, Western Board, Aero Zip, Treforest Zinc Printers and then ... Robinson's Flooring, then Stuart Singleham, then there was Plastics Engineering...

and so on for a further nine factories. It was 'chock a block' there.

This list reminds us of the fascinating array of goods that were manufactured. The variety and abundance of the often 'luxury' goods they produced seem at odds with the post-war period of austerity and rationing, but they were also aspirational and heralded a new exciting era. Women predominated in clothing and textile manufacturing. They made wool in Glyn Ceiriog woollen mill; rayon at Courtauld's, Flintshire; nylon at British Nylon Spinners, Pontypool; and elastic at Lastex Yarns, Hirwaun. Men's suits were produced at Hodges, Fforestfach, and high-end women's clothing at Alexon, Pontypridd, while many of the factories (such as Slimma's at Cardigan, Lampeter, Fishguard and Llandovery; St Margaret's Garment Factory, Aberbargoed; and John Stanton's, Llanelli) produced clothes to the stringent standards of Marks & Spencer. Across mid and north Wales, Laura Ashley's textile and clothing factories multiplied from Machynlleth to Carno to Gresford. Accessories abounded too, with gloves at Planet Gloves, Llantrisant; men's shoes at George Webb in Bridgend; and pocket watches at Anglo-Celtic (Smith's Industries) in Ystradgynlais. Underwear was manufactured at the Corset Factory in Caernarfon, and Berlei Bras and Kayser Bondor in south-east Wales. In Wrexham there was a sanitary towel factory and J. R. Freeman in Cardiff produced cigars. Toys, batteries, roller blinds, roller-skates, brushes, saucepan and tin components, washing machines, radios and televisions also flowed off the assembly lines. In Morris Motors, Llanelli, female workers coped with heavy manual work, making car radiators. Food and drink production was also represented at Avana Bakery, Cardiff; Smith's Crisps, Fforestfach; and Corona's, Porth, Rhondda.

This brief introduction indicates the complexity of this fascinating subject. It could have dissolved into a quagmire of facts and figures. It is rescued from this fate by the fact that this history is based upon the authentic, self-effacing female testimonies of the voices from the factory floor.

Chapter 1

Factory Girls

'Factory girls' may seem a patronising term for the thousands of women who worked in light industries during this period, but it does have some merit. After all, most of them were mere girls when they entered the workforce, and this is how they referred to themselves, even when they were older women. The 1944 Butler Education Act, stipulating that compulsory education should be extended to fifteen, was eventually enacted in 1947. In 1972, this was raised to sixteen. Most of the speakers, therefore, left school at fourteen or fifteen, some of them straight from primary education. Marguerite Barber (VSE072) testifies that she had attended only South Church Street School, Butetown, Cardiff, between the ages of four and fifteen when she left to work at Stamina Industrial Clothing nearby. Others attended the post-war secondary modern schools, but they often had no qualifications on leaving, as in the case of Anita Jeffery (VSE043). Illnesses, among them diphtheria (Vanda Williams, VSW044), rheumatic fever (Luana Dee, VSE015), polio (Jill Baker, VSE066), tuberculosis (Beryl Jones, VN027), an accident and ringworm (for sisters, Beryl Jones, VSW050 and Averil Berrell, VSW034), prevented many from sitting or passing the eleven-plus examination to gain the coveted entry into grammar school. There are several heartfelt stories, too, of parental deaths and illnesses, which exacerbated the situation and disrupted their education.

Poverty was a huge issue. Maureen Jones (VSE003) wanted to work because 'we were a poor family, I have to be honest', and Gwen Richardson (VSE018) 'simply wanted to earn money because of the poverty I had experienced through my childhood'. Some hated school and were pleased to leave. Vera Jones (VN042) 'couldn't leave school quick enough' to work in Courtauld's, and Sandra Brockley (VN050) says that, while her peers were crying, 'I clicked my heels together. Then I blossomed.' The older daughters in large families often had to sacrifice their ambitions for their siblings, especially if they were boys. Vicky Perfect (VN028) notes with some bitterness that her mother had been brought up in a different era, one where boys were expected to stay on and girls to go out to work; 'girls were undervalued', she says, 'my mum was a homemaker, that was her role.' Many years later, Vicky gained a diploma with the University of Oxford, in spite of this inauspicious beginning. Likewise, her mother's command – 'Once you're fifteen, you're looking for work' – landed Margaret Williams (VN014) in the Corset Factory, Caernarfon, in 1954. Neither of her parents would sign the form for Maureen Howard Boiarde (VSE031) to sit the eleven-plus examination, in spite of her teacher's protestations. As a possible future university candidate, 'It broke my heart. It absolutely broke my heart', she says.

Another cohort of factory workers interviewed did continue into secondary education. They passed their eleven-plus and even, in some cases, gained scholarships to pay for their grammar-school education. Gwynedd Lingard (VSE061), a very promising gymnast who later competed in two Olympic games, won the Caradog Wells scholarship, which kept her in her grammar-school uniform; however, she persuaded her mother to sign a form for her to leave at fifteen, so that she could pursue her gymnastic career. Commercial colleges, where girls could obtain qualifications in shorthand, typing, book-keeping and commerce, attracted many and the girls became clerks and secretaries in factories. Being a secretary was perceived to be 'a very glamorous job', claimed Jenny Sabine (VSW032), although she later qualified as a librarian. Others carried on into the sixth form and entered factory work as laboratory assistants.

Whatever their fate and, in spite of having aspirations to become hairdressers, nurses or teachers, it is striking how fatalistically many of them viewed their future destinies. As Patricia Howard (VSE029), who was born in the Rhondda and began working in Polikoff's at fifteen, states, rather whimsically, 'If I was a boy, I would have had to go to the pits. It's your gender that puts you where you are.' The interviewees express little regret or bitterness when they recount their early histories. Perhaps this is because they were schooled to regard work as only a break between school and marriage, and conditioned into believing that it mattered little how they spent the intervening years. Mavis Coxe (VN049) of Graesser's Salicylates, Sandycroft, comments that 'it wasn't a career really, it was a job', and this is echoed by Christine Jones (VN043) of Pilkington's, St Asaph: 'I didn't do it as a career … there's a difference between a job and a career, and I never had one of those.' Like so many other young girls of her class and generation, for Carol (VN021) *'work, marriage and having children was the plan.'* Mary Lynn Jones (VSW045) observed a similar pattern at Anglo-Celtic Watch Company (aka, Smith's Industries), Ystradgynlais, which was popularly known as 'Tick Tock':

> *Many of the factory girls started there at 15 and stayed for Mr Right to come along and, once the first baby came, finished and home. They were perfectly happy with this order of things. Some of them had been to the grammar school, able, but work was only a stop gap between leaving school and having children.*

Certainly, the 1960s' concept of teenage freedom and rebellion seems to have played little part in these young girls' lives.

Perhaps such attitudes relate directly to opinions about light industry in general in this period. Manufacturing was associated with the employment of female labour in particular, and thus considered of less value and significance than 'men's work'. Such disdain rubbed off on the female workforce itself. They were often depicted as rough and common; others 'looked down their noses at you' (Anon. VSW007). According to Patricia White (VSW058) of Lotery's, Newport, 'If you went to the clothing factory, you were working at the lower end of the scale … they didn't think you were worth much.' Several of the speakers were aware of this stigma. Such snobbish attitudes were sometimes shared by their parents. Phyllis Powell (VSE016)'s family 'had a fit' when she told them she would be working at Sobell's, Hirwaun, probably because factories had gained such a bad reputation during the war. Mair Richards (VSE025) recounts the drama in her home when she revealed she wanted to start work in Kayser Bondor, Merthyr Tydfil, in *c.* 1952. Her mother

> was ashes over the head you know ... 'Go to a factory to work – all those women with cigarettes and turbans on their heads (like the war, see!) ... You are not going to no factory.'

But, go she did, and Kayser Bondor turned out to be 'beautiful, gorgeous'. However, the balanced testimony of a former factory worker (VSE035) at Sidroy Mills, Barry, places these statements in their true context:

> Although women who worked in factories were regarded with disdain by some members of the public (those factory girls!), on the whole they were an intelligent lot, who, like my sister and I, were held back by the lack of opportunities available for women (and men) at the time.

It should be emphasised, however, that many of the speakers claimed that they were never subjected to such attitudes or prejudices. In their view, working in a factory was as valued a job as any other in their communities. Perhaps it is these conflicting views that illustrate most vividly how difficult it is to generalise about the interviewees' experiences. Although there are certainly common themes and threads, each story is unique and special. They are the stories of individuals – stories that sometimes fit, and sometimes do not fit, into the general picture.

One common pattern was for the girls to leave school on a Friday and start work on the following Monday. They were naive and easily impressed. Megan Owen (VN022) describes how she and her friend attended their interview at James Kaylor Compacts, Caernarfon, dressed in '*short white socks and ponytails, exactly like schoolchildren, and giggling childishly.*' Maisie Taylor (VSE011)'s mother accompanied her daughter to her interview at Peggy Anne's, Cardiff. During an interview at Courtauld's, Vera Jones (VN042) was subjected to be examined by a nit nurse, as if she was still in school. Interviews tended to be cursory affairs, more concerned with physical condition than mental ability or motor skills. Steinberg's, Treforest, organised medical and eye tests (VSE027) for their workers, and Meriel Leyden (VSW015) had to pass an eye test before being accepted on to the delicate watchmaking assembly line at 'Tick Tock'. Doreen Bridges (VSE054) refused a medical examination at Switchgear's, Blackwood: 'I was really annoyed ... I wasn't going to show my boobs to anybody! You had to be stripped to the waist', she protests. Though she was 'pulled over the coals' for her refusal, she was accepted into their workforce. By the mid-sixties, expectations had changed. Tryphena Jones (VSE074) had to undertake an IQ test to get a job at Freeman's Cigars and, by the early 1970s, Mettoy's, Fforestfach, had introduced an aptitude test. Margaret Morris (VSW043) was asked to 'put your washers, bolts in certain sequence, take it all off and do it with the other hand' within a certain time, to ensure that she was suitable for assembly line work at the factory. It is interesting to note that the companies on the Hirwaun Industrial Estate followed a code of conduct that they would not steal one another's workers, as Pat Howells (VSE082), a supervisor at Lastex Yarns, testifies. Prospective employees had to leave their previous employment before she would guarantee them a job on her watch.

Undoubtedly, family connections could secure a factory job. The Laura Ashley factories, especially at Carno, mid Wales, were run almost as family affairs in their early years. '*Someone from every house (in the village) worked in Ashley's*', says Margaret Humphreys (VN040); Gwlithyn Rowlands (VN013) could count her father, her mother and six siblings among her fellow workers there. At Morris Motors, 'a member of the family would be better than an outsider', claims one interviewee (Anon. VSW007) and Mary Lynn Jones (VSW045) says that the first thing they would be asked in an interview for 'Tick Tock' was to whom they were related.

Their first day in a factory made a huge impression upon these young workers. The noise, the rows of machines and the sheer number of workers left many frightened and bewildered. On her first day in 1966 at the 'massive, absolutely massive' de Havilland Aircraft factory, Broughton, Pat R. D. (VN051) was 'scared to death ... it was way out of my environment, my comfort zone.' There were very few like Maureen Howard Boiarde (VSE031), who revelled in this new experience. Starting in Polikoff's, in the Rhondda, in 1962 was, she says,

> Magical! ... I think the first thing you remember when you walk into such a big area is the noise. It's almost like a drone. ... There are hundreds of people talking ... and all the machines are going 'brrrr ... brrrr...' at high paces! ... the suits had to be steamed in the big presses ... 'sshhhhhhhh ... plonk!'

Training for a factory job took a fairly standard six weeks, though some picked up the skills 'by osmosis' (Sheila Hughes, VSE009). In Horrock's, they were taught to hand sew while they waited for a machine to become available, and in other clothing factories they had to master button-sewing, button-holing, and overlocking machines, among other skills. One laboratory assistant in Nettlefold's, Cardiff, was told she didn't need to know anything about chemistry to work there if she could follow a recipe (Janet Taylor, VSE041)! Indeed, Betty Metcalfe (VSE056) describes making calcium carbide at Kenfig Carbide's as 'like making cake. I followed the instructions – didn't always know why I was doing it'. Some enlightened factory bosses sent their young workers on day-release courses. Sarah Margaret Stanyer (VN047), for example, studied dressmaking at Flint Secondary Modern School on her away day from Courtauld's, while Lightning Zips, Waunarlwydd, insisted that their factory floor and office workers studied English, mathematics or shorthand and typing in the local technical college to improve their basic skills (Averil Berrell, VSW034). One interviewee (Evana Lloyd, VSW054) was sent from Felinfach Creamery, Ceredigion, to the Milk Marketing Board factory at Maelor to pass examinations to become a qualified laboratory analyst in the 1970s.

One aspect that galled several of the interviewees was that they could not become apprentices. In the fifties and sixties, this privilege was reserved for young male employees. There was a reluctance to train girls to become qualified toolmakers, draftswomen or engineers because, it was maintained, they would leave at the peak of their efficiency to get married or have children. One interviewee who was particularly frustrated by such male chauvinism was Eirlys Lewis (VSW061). After working in several other factories from 1964 onwards, it wasn't until *c.* 1976 that she was allowed to train as a machine operator at Vandervell Products, Cardiff. The men, she says, were quite willing for the girls to be given a chance, but they never really believed that they would be able to master the skills. She proved them wrong one day when one of the men had to appeal for her help. '*I let him stew for half an hour!*' she says, before taking all his work apart and getting the machine to work.

Not all the new factory workers were youngsters, of course. Many married women joined the workforce later in life, after raising a family. This often resulted in conflict between spouses and sexes. This is illustrated in singer Tom Jones's autobiography, *Over the Top and Back* (Michael Joseph, 2015), pp. 19–20:

> My mother was a housewife. She held a part-time job very briefly in a factory at the Treforest Trading Estate ... She pleaded with my father, who very reluctantly agreed with it. But one Saturday night we were queuing as a family to get into the pictures, and a man in the line – someone from the factory, whom my father didn't know – called out, 'Hello Freda.' The tone was no more than casually friendly, but it sent my father into a spin. It was the familiarity that got to him: not 'Mrs Woodward', but 'Freda'. My mum's adventures in the world of work ended shortly after.

Beryl Evans (VSW025)'s husband, Donald, worked in the Fisher & Ludlow car factory, Llanelli, but his sentiments echoed those above. '*Listen Beryl*', he said, '*I wouldn't like you to go to any factory to work*'; however, when he drowned while out fishing in the Loughor estuary in 1966, she had no option but to seek employment at INA Bearings nearby. One speaker (Anon. VSW007), who wanted to work at Morris Motors, Llanelli, overcame her husband's reluctance in a very practical way:

> He wasn't happy at all ... I didn't persuade him, I just told him I was going and that was it. And of course, when the money was coming in, he was quite happy.

Throughout this period, the main burden of doing the housework and organising childcare as well as working a full day still fell upon the working wife or even upon the young female worker in the household. 'A skivvy I was,' says Vanda Williams (VSW044) – for, when she returned from her night shift at Metal Box, Neath, she was expected to wash the floors on her hands and knees, beat the mats and clean the scullery before she could go to bed. Some speakers testify to the opposite and to the support they received from their husbands in regards to the housework, but they are in the minority. Once more, we are reminded that it is presumptuous to allow one to speak for the many or the many to speak for the few.

Factory Girls in the 1940s and 1950s

Right: Marian Bagshaw (right) and friend, Sylvia, at Wern Aluminium Works, Aberafan, in the mid-1940s. (Isabel Thomas (VSE040) Collection)

Below: Employees at Purma Camera Company, Hirwaun, *c.* 1946. (Ann Swindale Collection)

Workers at the flax factory, Milford Haven, *c.* 1946. (Poppy Griffiths (VSW039) Collection)

The workforce at 'Our Boys' pop factory, Briton Ferry, in the mid-1940s. (Isabel Thomas (VSE040) Collection)

Women making stockings at Bear Brand Hosiery, Corwen, 1947. (Geoff Charles Collection, National Library of Wales)

Customers checking the guillotine plate at Metal Box, Neath, 1949. (Skewen Industrial Heritage Association)

Employees at the Maesteg clothing factory work under the watchful eye of the supervisor, 1950. (Geoff Charles Collection, National Library of Wales)

Female employees at Berlei Bras, Ebbw Vale, 1951. (Geoff Charles Collection, National Library of Wales)

Machinists at St Margaret's sewing factory, Aberbargoed, Caerphilly, 1951. (Geoff Charles Collection, National Library of Wales)

Making toy hobby-horse walkers at Lines Bros/Tri-ang Toys, Merthyr Tydfil, 1951. (Geoff Charles Collection, National Library of Wales)

Right: Making a toy milk cart at Lines Bros/Tri-ang Toys, Merthyr Tydfil, 1951. (Geoff Charles Collection, National Library of Wales)

Below: Margaret Jones (VN024), standing, and her colleague, Megan making a doll's house at B. S. Bacon (Games) Ltd, Llanrwst, 1952. They retailed at around £7 each, says Vanda MacMillan (VN020), but she never bought one because 'working ... you used to get sick of them.' (Geoff Charles Collection, National Library of Wales)

Two workers at B. S. Bacon (Games) Ltd, Llanrwst, spraying a toy fort, 1952. (Geoff Charles Collection, National Library of Wales)

Processing tobacco leaves at J. R. Freeman Cigars, Cardiff, in the 1950s. (J. R. Freeman photographs@Gallaher Ltd)

Right: Operators washing tobacco leaves at J. R. Freeman Cigars, Cardiff, in the 1950s. (J. R. Freeman photographs@ Gallaher Ltd)

Below: Nesta Davies (VN025) at her loom in Johnson's Fabrics, Wrexham, in the early 1950s. She met her husband in the factory. He cleaned her loom, and 'I had the cleanest loom in the place', she says. (Nesta Davies Collection)

Workers at their machines at the Anglo-Celtic Watch Company (locally called 'Tick Tock'), Ystradgynlais, in the 1950s. (Glamorgan Archives)

Workers at Hunt & Partners (box factory), Bridgend, in the early 1950s. (Mrs P. P. (VSE036) Collection)

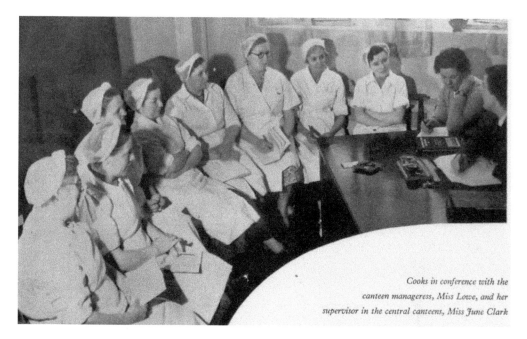

Cooks in conference with the canteen manageress, Miss Lowe, and her supervisor in the central canteens at British Nylon Spinners, Pontypool, in 1953. (From the *British Nylon Spinners* magazine, April 1953, p. 2. Copyright of Pontypool Museum)

Customers checking the quality of jar top ends at Metal Box, Neath, 1956. (Skewen Industrial Heritage Association)

Above: Feeding tinplate on the high-speed 314 press, at Metal Box, Neath, in the 1950s. (Skewen Industrial Heritage Association)

Left: Rosie Jones (VN032) and Mair Griffiths (VN033) in the laboratory at Llangefni Creamery, Anglesey, 1950s. (Rosie Jones Collection)

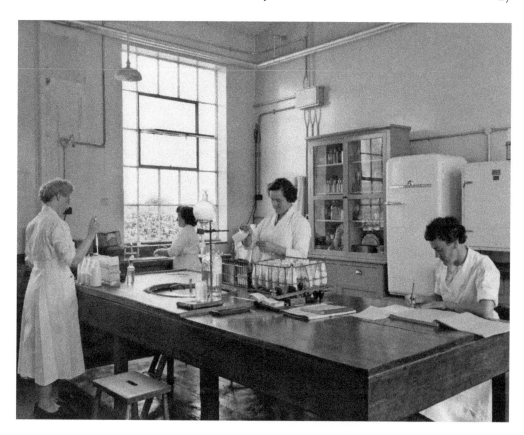

Above: Mair Griffiths (VN033) with colleagues in the laboratory at Llangefni Creamery, Anglesey, in the 1950s. (Mair Griffiths Collection)

Right: Nanette Lloyd (VSW004), right, and Glenda Lewis (VSW003), third from left, in John Patterson's (Fairweather Works), Pont-henri, Carmarthenshire, in the mid-1950s. They made tablecloths; Nanette is wearing a remnant of material as an apron to protect her clothes. (Nanette Lloyd Collection)

Above: Corgi Hosiery workers, Ammanford, early 1950s. (Corgi Hosiery)

Left: Marge Evans (VSE052) sitting at the Westminster machine, making coils at Sobell's, Hirwaun, in the 1950s. (Marge Evans Collection)

Mair Matthewson (VSW046) packing open tops (750 tops a minute) in Metal Box, Neath, in the 1950s. *'You couldn't touch anything without your gloves,'* she says. (Mair Matthewson Collection)

Stuffing teddy bears at H. G. Stone's, Pontypool, in the 1950s. 'The women on the teddy bear aisle got more than the rest of us. They thought that was a better job than on the other soft toys,' says Mary Farr (VSE010). Violet Skillern (VSE083) sits fourth from the front. (Violet Skillern Collection)

Operating a machine at Lines Bros/Tri-ang Toys, Merthyr Tydfil, in the 1950s. (Marjorie Collins (VSE026) Collection)

Ann Davies (VSE032) at her machine in J. R. Freeman Cigars, Cardiff, 1957. (Ann Davies Collection)

Workers outside Lines Bros/Tri-ang Toys, Merthyr Tydfil, in the 1950s; Marjorie Collins (VSE026) is third from the left. (Marjorie Collins Collection)

Averil Matthews (later Berrell, VSW034), standing, and Brenda Hooper as young clerks using a Remington typewriter in the office in Lightning Zips, Waunarlwydd, Swansea, in 1955. (Averil Berrell Collection)

Nan and Meiryl James (VSW053) test milk on the deck at Felinfach Creamery, Ceredigion, in *c.* 1959. (Meiryl James Collection)

Chapter 2

Work, Work, Work

Factories were deeply gendered workplaces. Almost invariably, the bosses, the managers and those in authority, such as foremen, were men. On the shop floor, almost all the women questioned felt that there was 'men's work' and 'women's work'. In the clothing trade, for example, men did the cutting out and pressing, working with the larger machines, and they were also the mechanics; meanwhile, the women were the machinists on the assembly line itself. Yet, women also riveted, soldered, drilled, deburred and countersank to make washing machines at Hotpoint, Llandudno, and Hoover's, Merthyr Tydfil; jewellery at Attwood & Sawyer's, Porthcawl; or switches for Switchgear, Pontllanfraith. They also drove forklift trucks and overhead gantries and cranes. Sandra Brockley (VN050) describes how, as 'slingers', they managed to add a feminine touch, even when working an overhead crane to suspend metal into a trichloroethylene bath at de Havilland, Broughton:

> You're going to be up there for a couple of hours, so you'd take your lunch with you. If you wanted baked beans ... they'd tie a piece of wire, dip it in the bath and leave it there for a little while and then you'd have hot beans ... the girls, because they knew they were stuck there ... took their knitting with them.

One of the most difficult aspects of the interviewing process was to understand the exact nature of the interviewees' work. In Courtauld's, the women describe spinning, cake washing, cake wrapping, coning and cheesing – terms that are unfathomable to the uninitiated. It is interesting to note too that Welsh-speaking interviewees peppered their descriptions of their tasks with technical English terms, although they weren't always aware they were doing so.

There was 'an inexorable quality' to the non-stop production of the assembly line, according to Annest Wiliam (VSW031), who was a temporary student worker at Mettoy's, Fforestfach in the 1950s. Even in the smaller factories, your nose would be kept to the grindstone, as at the Sweets' Factory, St Asaph, where 'you'd wrap and wrap as fast as you can and fill your boxes as fast as you can' (Joyce Edwards, VN045), or at the cockle factory, Caernarfon, where, according to Dilys Wyn Jones (VN006), '*God we worked hard, we were like navvies.*' Irene Hughes (VSE024), who fed the male cutters at Kayser Bondor, Merthyr Tydfil, for over thirty years, called it 'Chinese labour':

The men ... used to earn the money and Mair (VSE025) and I were the donkeys lifting heavy rolls and running up and down laying them, and they used to go round on the band knife. They were well away. We weren't though.

Irene was infuriated at having to sit under a sign that quoted Sir Stafford Crisp: 'Output is the key to prosperity'. 'I could have ripped it off the wall!' she proclaimed. The motto that adorned the wall at British Nylon Spinners, Pontypool, must have been especially galling to the female workers there: 'The spear-head of the works is the trained man on the shop floor'[5]. The sheer hard labour on the factory floor was one aspect that impressed temporary holiday workers, who were interviewed for the project for their valuable, alternative interpretations as 'outsiders'. Packets of crisps flew past Jenny Kendrick (VSE020) at Smith's Crisps, Fforestfach, and the nail varnish bottle tops landed in Susan Leyshon (VSE063)'s lap because she couldn't close the lids fast enough to keep up with her line at Revlon, Maesteg. Edith Williams (VSE062) tells an amusing tale about her experience of the non-stop production of cheese balls at OP Chocolates, Merthyr Tydfil:

> One day the elastic went in my trousers and they started slipping down, and I couldn't stop, 'cos if I did stop the biscuits would be all over the floor. So I was shouting 'Help, help!' for somebody to come to take over.

Enid Thomas (VSW066), an industrial nurse who served at Fisher & Ludlow, Llanelli, voiced her professional opinion of some factory managers' attitudes: '*Some expected too much, especially of the women ... They pushed them, pushed them, wanting them to do more.*' It is striking in this context that most of the photographs taken in factories during this period show rows upon rows of women at their individual machines or at conveyor belts, heads down, concentrating on their work, in a 'very, very regimented working environment', to quote Lynfa Macer (VSE065).

Women, it was believed, were suited for dull, repetitive work, and particularly for tasks that demanded manual dexterity and patience. This, according to Maureen Howard Boiarde (VSE031), was a further manifestation and extension of their domestic labour:

> Women could sit and do the same job over and over again 'cos that's what every housewife does every day – she polishes the floor, washes the same dishes, cleans the same carpets. Women are great at that.

The women often colluded with this view. You had to have '*a light, small hand*' (Catherine Evans, VSW016) to cope with the tiny balance wheels that went into pocket watches at 'Tick Tock', for example.

Attitudes to the work varied immensely. Some of the speakers considered it to be very boring and monotonous. For example, Nan Morse (VSW017) describes her short time at Alan Paine, Ammanford, thus: '*you sat in your seat and did the same thing from morning to evening. You went for your lunch, and you came back and did the same thing ... I felt as if I was in a box ... It was like a jail sentence. It was horrible.*' Likewise, Anne Amblin (VSE022) was not impressed with the bulb-loading she was expected to do at

Thorn Electrical, Merthyr Tydfil, because 'you could train a donkey to do it after a while'. However, this was not the consensus of opinion held by the majority of the interviewees. In the clothing trade, there was the mastery of different machines, being skilled enough to become a 'floater' able to take over anywhere on the shop floor as required, or completing whole garments without having work returned to you by those on inspection; this gave the workers a sense of achievement and pride. Top machinists, explains Sylvia Howell (VSW062), who worked in John Stanton's, Llanelli, fitted collars, yokes and cuffs, as opposed to just overlocking seams or finishing hems. At the sewing factory, Denbigh, Eira Richards (VN046) was proud of the fact that she was expected to make whole garments – cutting, sewing and pressing them herself. Each new style or different-coloured garment presented its own challenges for workers at Laura Ashley's: '*Perhaps ten blue, six red and four green dresses were on order ... you would have to change the cottons all the time,*' says Olive Jones (VN039). Marks & Spencer were particularly demanding taskmasters, even determining how many stitches per inch they should use (Eileen Davies, VSW026), and, when their inspectors visited Slimma's in Cardigan, 'everybody was supposed to be on their best behaviour' (Anon., VSW024).

Several of the speakers identified with their specialist machines. Anita Jeffery (VSE043) felt quite important when she was promoted to making sleeves on the German Pfaff machine in Polikoff's, the Rhondda; meanwhile, because she was one of the best workers at Corgi's Ammanford, Patricia Murray (VSW019) was given a bigger sewing machine, which enabled her to do more and win more money. As she points out, her work, involving using different gauges of wool, was highly skilled and 'very, very interesting'. It is fascinating to hear some interviewees speak with great affection of their particular machines and equipment. At Lewis & Tylor, Gripoly Mills, Cardiff:

> There was one loom that everybody loved. I don't know what it was ... we used to fight to try and get on this one loom. You seemed to have a better quality belt off it. But I always wanted to keep my own wires and my own rods, my own beater, my own needle and my own tape. (Jill Williams, VSE051)

Marge Evans (VSE052) describes a similar experience when she was making television sets in the 1950s at Sobell's, Hirwaun. She was moved to operate the big Westminster machine, which could make ten coils at a time:

> In the beginning it was very frightening, but then we got that we used to run our hands over it when it was going full speed ... because it was part of us, wasn't it?

These women were proud of their own skills and of the goods they manufactured. Two workers at Morris Motors, Llanelli, echo these sentiments: '*You had pride in your work – when you saw something you had done and you knew that it was perfect*' (Patricia Lewis, VSW028); 'I used to take pride in my work and the quicker I could go and do it perfect, the better I felt' (Yvonne Bradley, VSW063).

In most of the larger factories, the women 'worked to the hooter'. Although Patricia Howard (VSE029) was accustomed to colliery hooters in the Rhondda, when her own life became 'automated to a buzzer', it became personal. Clocking in and clocking out were important factory rituals. Every worker had a number: '*I remember mine in Austin Taylor's ... 3359 and, if we were a minute after eight o'clock clocking in, we lost half an hour's pay*', declared Sandra Owen (VN054). Many had long forgotten their factory numbers, yet they all remembered the penalty meted out to latecomers, although the exact timings varied from factory to factory. Marilyn Hankin (VN048) of Glaesser's Salicylates noted that pay would be docked if workers clocked out a minute early too. At Hotpoint, Llandudno Junction, 'There was a line that they had to stand behind in order to go out at the end of the day and the managers would watch that nobody stepped over it before the official leaving time and the bell went' (Margaret Evans, VN037). Clocking in or out on someone else's behalf could bring instant dismissal. Mary Evans (VN008), who worked making compacts in Caernarfon, got away with it several times, she says, but saw '*lots of others having "instants"*.' In factories where they had to clock in and out for dinner and, in some cases, tea breaks, the management had even greater control over its workforce. Gwen Richardson (VSE018) found her own way of coping with the short tea break at Planet Gloves, Llantrisant:

> The canteen was on the ground floor, you worked on the top floor, you had ten minutes to come down three flights of stairs. It's funny but if I'm faced with a long flight of stairs even now, I'll still ... hold onto the sides and slide down ... it was the fastest way so that you'd get in the queue for your tea and toast ... because you had to be back on your machine by the time the buzzer went for the end of break.

Smith's Crisps was 'done by whistles' too, according to Era Francis (VSE075). They had an hour for dinner (or half an hour in some factories), and they had to queue for food and queue for tea, wash their hands and be back at their machines before they were switched on again. Not all factories were as regimented, of course. Luana Dee (VSE015) describes how the young girls and boys at Tri-ang Toys, Merthyr Tydfil, used to bunk off after being paid on Friday afternoons to 'chill out' in Cyfarthfa Park. Then, 'we'd go back to work on Monday and nobody would ever say anything about it.' It is interesting to contrast these factory-floor testimonies with the experiences of laboratory staff at British Nylon Spinners, Pontypool, where they were allowed ten minutes before lunch and ten minutes before going home 'to smarten up' (Sheila Hughes, VSE009). Altogether, as the testimony above shows, and in spite of some contrary comments, none of the 200 female workers interviewed for this project would be able to identify with singer Max Boyce's satirically mischievous portrait of factory life in his popular 1971 song, 'Duw it's Hard', based ostensibly on his own experiences as an apprentice electrician at Metal Box, Neath:

> But I know the local magistrate, she's got a job for me
> Though it's only counting buttons in a local factory,
> We get coffee breaks and coffee breaks, coffee breaks and tea
> And I know those dusty mines have seen the last of me.[6]

Several of the very large factories had segregated canteens, which emphasised once again the hierarchical structure of the factory environment. In AB Metals, Abercynon, there were three canteens: for the factory floor, for staff and for the directors (Sylvia Reardon, VSE006). Glenys Rees (VSW059) had 'a big issue' with such ghettoization of factory workers in the canteen at Louis Marx toy factory, Fforestfach, for 'the canteen staff put tablecloths for the office workers but not for us ... I thought that was awful', she says. This differentiation was even more marked at Freeman's Cigars, Cardiff, according to Tryphena Jones (VSE074), who worked there between 1964 and 1977:

> The floor workers were in the big canteen; and the charge-hands, the foreman had another ... they were served, and the office workers had another room, and management another room, which was silver service ... the normal workforce had to queue up in the canteen.

To exacerbate the situation, they also had different food served in the layered canteens. Although there were canteens in many of the factories at this time, they varied greatly, and most of the women would rather bring their own sandwiches 'in a snapping tin', as in the case of Nesta Davies (VN025), to eat on their lunch breaks, so as not to waste their precious, hard-won earnings on food.

More Factory Views

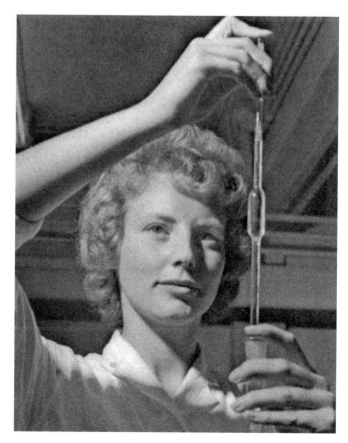

Research laboratory assistant, Pamela Stevens, carrying out part of a trace-metal analysis on polymer at British Nylon Spinners, Pontypool, *c.* 1960. (From *British Nylon Spinners* magazine, winter 1960, p. 46. Copyright of Pontypool Museum)

Debbie Edwards and Marian Gregson, laboratory workers at Pont Llanio Creamery, Ceredigion, *c.* 1960, on a forklift truck in the factory yard. (Mair Richards (VSW057) Collection)

Megan Owen (or Megan Bach, VN022) at her machine in James Kaylor Compacts, Caernarfon, in the 1960s. (Dafydd Llewelyn Collection)

Mary Evans (VN008) and colleague in James Kaylor Compacts, Caernarfon, 1960s: '*I loved it there … There were some rough and ready girls there, but I really liked them … I could listen to their stories; things I wasn't allowed to hear at home. God!*' she says. (Dafydd Llewelyn Collection)

Above: Operators at AB
Electronics, Abercynon,
in the 1950s. (Paul
Norton photographer;
Glamorgan Archives)

Left: Operators at
their machines at AB
Electronics, Abercynon,
in the 1950s. (Paul
Norton photographer;
Glamorgan Archives)

Female employees in the Rotor Board Assembly Tuner Division at AB Electonics, Abercynon, in the 1950s. (Paul Norton photographer, Glamorgan Archives)

Workers at Ferodo car factory, Caernarfon, in the mid-1960s. (Geoff Charles Collection, National Library of Wales)

'Working to the hooter'. The clocking in and out clock at Courtauld's, Flint. (Vicky Perfect (VN028) Collection)

Marion Jones (VSE028) and Kate Collins at work in Hoover's, Merthyr Tydfil, *c.* 1965. (Marion Jones Collection)

Above: Barbara Vaughan at her machine in Hoover's, Merthyr Tydfil, *c.* 1965. (Marion Jones (VSE028) Collection)

Right: A female apprentice at South Wales Switchgear, Pontllanfraith, in the 1960s. Apprenticeships had been reserved traditionally for male employees, so female apprentices were unusual. (Hawker Siddeley Switchgear)

Above: Eira John (VSW048) and colleagues in the warehouse packing department at Mettoy's toy factory, Fforestfach, Swansea, in the 1960s. (Eira John Collection)

Left: Female workers at Thorn Electrical Industries, Merthyr Tydfil, in the 1960s. (oldmerthyr.com)

Workers at the Chloride (battery-making) factory, Pont-henri, Carmarthenshire, in the late 1960s. (Pontyberem Historical Society)

Workers at the Chloride factory, Pont-henri, Carmarthenshire, in the late 1960s. (Pontyberem Historical Society)

Susie Jones (VN016) and fellow male workers in the laboratory department, Cooke's Explosives, Penrhyndeudraeth. '*I didn't have any higher or college education but working in the Works was an experience and an eye-opener,*' says Susie. (Susie Jones Collection)

The cigar-making machines at J. R. Freeman Cigars, Cardiff, in the 1960s–70s. (J. R. Freeman photographs@Gallaher Ltd)

Right: Audrey Gray (VSE050) in the laboratory at Johnson & Johnson, Pengam, Caerphilly, late 1960s. (Audrey Gray Collection)

Below: Draftsmen and draftswomen in South Wales Switchgear, Pontllanfraith, late 1960s. (Hawker Siddeley Switchgear)

Particia Ridd (VSW041),
with her colleague at
Corona pop factory,
Swansea, in the late 1960s.
(Particia Ridd Collection)

Gaynor Hughes (VN029) at
her machine in Courtauld's,
Aber, Flint, *c.* 1970. (Gaynor
Hughes Collection)

Machinists at Lotery's, Newport, 1971. (Patricia White (VSW058) Collection)

Jill Williams (VSE051) training a young worker at Lewis & Tylor (Gripoly Mills), Cardiff, in the mid-1970s. (Jill Williams Collection)

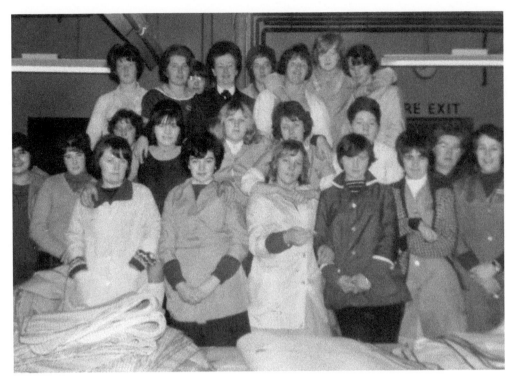

Group of workers, including Augusta Davies (VSW011) in the middle row, fourth from left, at Cardwell clothing factory, Lampeter, in the mid-1970s. (Augusta Davies Collection)

Employees at their machines at Hotpoint, Llandudno Junction, in the late 1970s. (Kathy Smith (VN023) Collection)

Workers at Alan Paine knitwear factory, Ammanford, celebrating winning the Queen's Award for Export, with Patricia Murray (VSW019) seated in the middle. (Richard Firstbrook, Llandeilo)

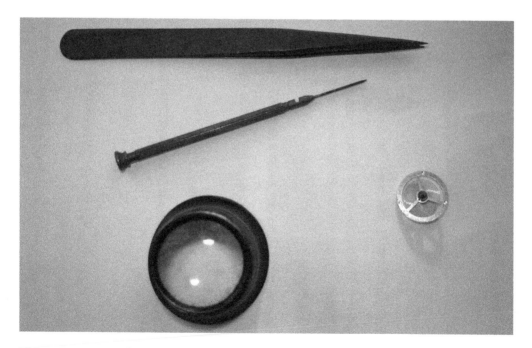

The tweezers, screwdriver and eyeglass used by Meriel Leyden (VSW015) to assemble watches with balance springs at the Anglo-Celtic Watch Company ('Tick Tock'), Ystradgynlais, from 1955 to 1980. To tease new workers, they used to colour the rim of the eyeglass with a blue pencil, which would leave a blue mark around their eyes for the rest of the day. (*Archif Menywod Cymru* Women's Archive of Wales)

Packing the cut-out parts in Arlee Textiles (part of Louis Edwards, Maesteg), Neath Abbey, in the mid-1970s. Each conveyor belt was capable of packing 2,000 garments a day, subject to the number of parts per garment. (Skewen Industrial Heritage Association)

The workers outside Vandervell Products, Cardiff, in the 1970s. The factory produced parts for lorries and cars. (Eirlys Lewis (VSW061) Collection)

Yvonne Bradley (VSW063) in Morris Motors, Felinfoel, Llanelli, in the mid-1970s. (Yvonne Bradley Collection)

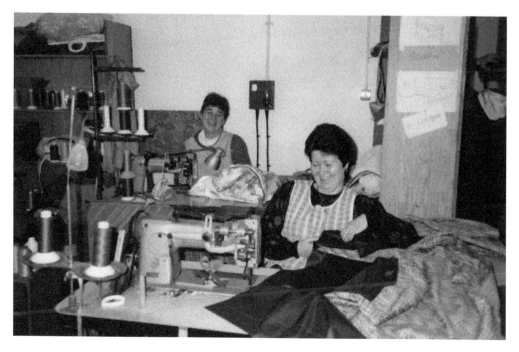

Caroline Aylward (VSE055) in Christie-Tyler, Glyncorrwg, Neath Port Talbot, in the late 1970s. (Caroline Aylward Collection)

Chapter 3

Earning Their Crust

Trying to record and fathom the intricacies of factory wages and pay over the timespan of a quarter of a century has proved to be a minefield of contradictions and anomalies. There are so many variants, including the actual date, the ages of workers, the nature of their work, decimalisation, the hours they worked and, most taxingly, the way their wages were calculated.

Working hours varied from factory to factory and also changed considerably within the timeframe covered by this project. It was common to start at 7.30 or 8 a.m. and work until 5.30 or 6 p.m. Joyce Evans (VSW027) notes that, when she started in 'Tick Tock' in 1947, she was on a 48-hour week; however, when she left in 1988, this had been reduced to 39 hours. Others describe horrendously long hours, especially when they had no choice but to do overtime. Sandra Owen (VN054), of Austin Taylor's, got up at four in the morning to walk to Bethesda for the 6.00 a.m. to 2.00 p.m. shift; then, because of the penalty shift to complete orders by the end of the month, she often found herself working on until six in the evening, going home and returning at midnight to work until six in the morning. However, most of the women interviewed were happy to work overtime – a few hours extra in the evenings or 4 hours on Saturday mornings paid at time and a half, or at double time on Sunday, could boost their meagre incomes significantly and help their factory to meet customers' demands. Extra shifts were introduced as necessary to satisfy these demands. Revlon, Maesteg, brought in a 'mothers' shift' from 10 a.m. until 2 p.m. in the 1960s (Rosalind Catton, VSE060), while Laura Ashley was considered a very enlightened employer because she would not allow any working mother to work after 2.45 p.m., so that they could fetch their children from school. 'She was all for the mother', claims Mo Lewis (VN002). The Factories (Evening Employment) Act of 1950 introduced the 'twilight shift' to encourage the mothers of young children back into work. At 'Tick Tock', this early evening swing shift brought many more married women into the workplace. Moira Morris (VSW001) found this very helpful when her collier husband was out on strike and money was in short supply. Joyce Evans notes that friction arose with the day shift workers in the clock section of this factory as the swing shift workers were hitting their targets. Shift work was a controversial issue in several of these factories, and this will be explored further in Chapter 5.

Although it's almost impossible to generalise about the wages female workers received, certain patterns are discernible. In the immediate post-war period of austerity, from the late 1940s to the early '50s, a fourteen- to fifteen-year-old girl would earn between £1 and £2 a week in the factories. These wages were higher than those earned in a shop, although working in Woolworths was considered quite prestigious, and its sweet counter was particularly attractive to Ann Owen (VSE005). The interviewees remember receiving their first wage packet: 'It was thrilling, exciting really', says Anne Amblin (VSE022). The seemingly low wages reflect the fact that often these young girls began at the bottom of the ladder as untrained factory workers. Beti Davies's (VN009) first wage at Glyn Ceiriog Woollen Mill in 1944 was 5s a week, which *'didn't seem much at the time'*, but, when she left six years later, it had risen to £1 3s 6d and was nearer the norm. Many of these factories were owned by Jewish families – e.g., Lotery's, Newport; Sidroy Mills, Barry; and Mettoy's, Fforestfach – and there was a perception that they were keen on very young workers as they could pay them less (Cynthia Rix, VSW052). Between the second half of the 1950s and the duration of the 1960s, wages seem to have crept up gradually, so that Enid Jones (VN052) was earning £7 a week in 1957 at Ferranti's, Bangor, and after ten years' service Margaret Young (VSW065) was earning £12 14s 4d a week at Corgi Hosiery, Ammanford, in 1969. Certain employers were considered very good payers. Marion Jones (VSE028) felt like a 'millionaire' when she received £10 in her pay packet at Hoover's, Merthyr Tydfil, in 1963, especially as she also had weekly, monthly and Christmas bonuses. The view that J. R. Freeman's Cigars was one of the best payers is supported by the pay packets that Frances Francis (VSE023) has kept from her time at the factory. In July 1964 her basic wage was £7 17s 9d for a 42-hour week but, by April 1966, she was earning £9 1s 6d for a 40-hour week. Kayser Bondor, Merthyr Tydfil, had a similar reputation. 'We were highly paid in those days', reminiscences Edith Williams (VSE062), and Irene Hughes's (VSE024) pay slips from 1949 and 1974 confirm this view. In 1949 Irene earned £1 15s gross a week but, by 1974, she was on £31 13s.

Such bare facts and figure must be seen in their historical perspective. Throughout, female factory workers were considered of secondary significance to their male colleagues. Their work was classified as unskilled and was therefore low waged, no matter how skilled they actually were as competent top machinists or proficient assembly-line workers. Unfortunately, this is also how most of the women saw themselves and, consequently, they often undervalued their own work (Rosie Jones, VN032 and Anon, VSW006). The details of pay structures printed in the Milk Marketing Board's Staff Handbook in the early 1960s illustrate this. A fifteen-year-old male factory-floor worker would earn £5 12s, while a female of the same age would earn £4 17s; a twenty-one-year-old male would earn £11 16s and a female £9 4s 6d. Even in the laboratory at Kenfig Carbide, Betty Metcalfe (VSE056) says that the 'boys' were always paid more than the girls. 'We didn't really like it', she admits, 'but it was such a wonderful firm to work for.' Similarly, Doreen Bridges (VSE054) considered her job spot welding ironing boards and kitchen stools at Golmet's, Pontllanfraith, a skilled one, but 'some of the men were only brushing floors and they were getting much more than what we were getting, and we were on machines. I didn't like that.' According to Tryphena Jones (VSE074), the mechanics on the shop floor in Freeman's Cigars, Cardiff, would just 'float around all day with a screwdriver in their pocket, and just go sitting down waiting for something to happen', and yet they earned a lot more money than the female assembly-line workers, who could in fact fix their own machines when necessary and dispense with the mechanics' 'expertise'. When questioned about the parity of wages between male and female workers at AB Metals, Abercynon, Sylvia Reardon's (VSE006) answer was typically forthright:

> Parity with men? There was no parity. You must be joking ... you knew the men were on at least 75 per cent more than you. And the old argument was always, they've got families to keep, and it was as if the whole of the male population classed a working woman, whatever her personal circumstances, as working for, in inverted commas, pin money, whereas a lot of these women, particularly single women like myself, were working to support yourselves.

In this respect, it is interesting to reflect upon Michele Ryan's (VSE070) comments on these attitudes. As a temporary holiday worker at the Glass Works, Cardiff in 1969/70, she considered the works as a residue of the war years; the men had 'come back, they needed their pride, they needed their sense of being the bread winner ... the head of the household ... a wife could only help out.' Several speakers talk of the disparity between their and their husbands' wages. Even though Anita Jeffery (VSE043) was a skilled machinist at Polikoff's, Rhondda, she received a lower wage than her husband, who was a packer in the same factory. On the other hand, several of the speakers note with pride that they earned more than their husbands during this period. May Lewis (VSW002) earned a higher wage in Morris Motors, Llanelli, than her collier husband in the 1940s, while a machinist (VSE068) in Louis Edwards, Maesteg, earned double her underground-miner husband's pay in the 1960s.

As the discussions above suggest, there was a considerable difference in attitudes towards the wages of young single girls, who did not get a full wage until they were twenty-one, and those of married women, whether working full- or part-time. Almost universally the young girls would turn their pay packets over unopened to their mothers on pay-day Friday. Cynthia Rix (VSW052) describes how her mother 'met her by the bus, holding out her hand for the wage packet'. Young male workers and husbands also handed over their earnings in this way in many working class households (Margaret Williams, VN014). In return, most received pocket money, while others only got the bare minimum to cover bus fares and dinner money. One speaker (VSW034) describes herself as being '*as poor as a church mouse*' because her mother didn't allow any extra money from her own pay packet to buy clothes. Rosie Jones (Llangefni Creamery, VN032) spent her allowance of 10s a week on bus fares, a packet of chips and going to the pictures in Bangor every Saturday night. This kind of parental arrangement was the norm and, in some cases, it carried on until the workers got married. The usual pattern, however, was that, once they reached eighteen (Kathleen Matthias, VSE069) or twenty-one (Jill Baker, VSE066), they paid board and lodging only. Some mothers objected to this: Kitty Jones (VSW056)'s mother went 'ballistic' when she approached her with the new terms and Era Francis (VSE075) had 'to strike a bargain' with her mother to allow her to keep a fairer portion of her pay to save for her impending marriage. Maureen Howard Boiarde (VSE031) says that, if mothers didn't put their daughters or sons on board and lodge, it was deemed they 'were stealing from them. Bartering it was! It was like a rite of passage'.

The wages earned by married female factory workers were often called 'pin money', a derogatory term for such hard-earned money. Many of the workers themselves devalued their work, and felt that they were merely supplementing the main family income, earned by the main breadwinner in the home. In fact, these female workers made a very significant financial contribution towards raising their families out of post-war austerity into the relative prosperity of the 1960s and 1970s. In the early years, they describe how they made television sets, telephones, cars, toys, and clothes they could never afford. Beryl Roberts (VSE033) made parts of television sets at Masteradio, Treforest, in the early 1950s, and 'often wondered whether they worked or not', but couldn't afford a set herself until she got married in 1955. Yet, by sheer hard work and considerable sacrifice of personal ambition, many of these women succeeded in realising some of their dreams. Their tenacity provided their families and especially their children with a better way of life and more opportunities, through school holidays, university education and other 'luxuries', that they themselves had not been fortunate enough to experience. It is a disservice to them to class this contribution as mere pin money.

The working environment in many factories was governed by reaching set targets. These would be determined by the time-and-motion officials, whose presence dogged the factory floor. In the clothing trade, says Maureen Williams (VSE030),

> Each job would have a different cost; for example, sewing a straight seam would be priced differently to ... a gusset. Sewing black garments was paid at a premium as it was considered much more difficult.

Former workers at Sidroy Mills, Barry, (VSE035) suggest that, when they were being watched by time-and-motion officials, the selected machinists would feel under pressure to work as fast as they could, 'ignoring the possibility of toilet breaks or even to wipe noses' and that this conspired to keep wages low. However, most of the interviewees responded more pragmatically to this challenge. Maisie Taylor (VSE011), who worked at Horrock's, Cardiff, was urged by her fellow workers to slow down when being timed in order to keep the targets low, and Tryphena Jones (VSE074) describes how she would work exactly to the book, checking insignificant details, during time-and-motion studies at J. R. Freeman's Cigars, Cardiff. Patricia White (VSW058) summed up the effect such scrutiny has had on her life and work:

> If the time-and-motion people were setting a target and timing somebody, you would try and stretch the task, (but) when performing it for your own purposes you would minimise movements in order to maximise your earnings. The whole thing was to cut down on any unnecessary movement. That's the only way you could beat the timings. You could not be fiddling around with scissors ... You had to be concise with all the movements, because you were talking about cutting seconds off ... It's something that's affected the whole of my life since ... It makes you very efficient.

As a supervisor, Pat Howells (VSE082) readily admits that the workers could be very crafty when time-and-motion studies took place!

In spite of such shrewdness, the targets set often seem immensely challenging. The whole of the production line would be involved in meeting some targets: for example, producing 500 washing machines a day at Hotpoint, Llandudno Junction (Kathy Smith, VN023); making seventy-five pyjamas every half an hour at Aykroyds & Son, Bala

(Cath Parry, VN001); packing 750 open tops per minute in teams of four at Metal Box, Neath (Mair Matthewson, VSW046); or labelling and packing 800 tins of crisps a day at Smith's, Fforestfach (Monica Walters, VSW049). The workers knew they would be letting the rest of their line down if they did not keep up with their targets. Others had individual targets to meet. Helena Gregson (VSW009) was expected to hem 400 trouser hems every 2 hours at Slimma's, Lampeter, and as she says:

> You didn't have time to talk to people ... in work. Oh no, you couldn't go off your machine to gossip or your targets would be out of the window ... Anyone who had worked in a factory like Slimma's – they weren't afraid of work.

If and when the required targets had been met, some workers could work on to earn a bonus. Sylvia Howell (VSW062), John Stanton's, Llanelli, describes how,

> after you had done your target first, from there on in the week you could go on bonus – so much per ten garments. If you had a good run, the machines didn't break down, or there was no problem with the patterning, and the supply was coming through, you could actually bump your £8 up to £2 or £3 extra – to £11 before stoppages.

This was not always practical. Luana Dee (VSE015) was consistently earning bonuses as a fast worker with good eyesight at Berlei Bras, Merthyr Tydfil, and consequently was moved onto the black bra line. As she explains 'it wasn't easy to sew black on black', so there were lots of mistakes. When she asked through her union for a lower target, she was moved to another part of the factory. Sylvia Reardon (VSE006) recounts a bizarre tale about the pressure of reaching targets in the invoicing department at AB Metals, Abercynon. Her mother had acquired a supply of Dexedrine tablets from a doctor's surgery, without realising how dangerous they could be. At the end of every month the office staff had to work 'like slaves', so Sylvia gave them a tablet each to boost production:

> Good god, there were sparks coming out of the typewriters ... We were like maniacs ... You could do a full day's work, go home, strip the wallpaper off, repaper and do the washing and ironing and turn up for work the next day ... No wonder we thought ABs was wonderful!

Piecework, where workers were paid according to how many goods they produced, was another system popular with many employers in this period. It was also popular, of course, with fast and accurate workers. Gwen Richardson (VSE018) was adept at managing her time on piecework at Planet Gloves, Llantrisant: 'I was quite a fast worker and I always only took 15 minutes for my lunch and then I'd go back and sit at my machine and do all the cutting and things', she says. There was a general consensus of opinion that piecework could be extremely demanding, although lucrative, even when workers were healthy and efficient. It certainly did not favour the disabled; Gwen Evans (VSW014), who carried a disabled card because of a weak arm, found piecework very difficult when making car radiators at Morris Motors, Llanelli. When a worker at Cooke's Explosives, Penrhyndeudraeth, sought the doctor's advice for a thyroid problem, his response was unequivocal: 'They are making machines out of you instead of human beings ... Piecework is wrong. I condemn it' (Blodwen Owen, VN026).

Office clerks calculated these various, complicated wage structures by means of tickets or timesheets, and by using their comptometer machines, using the Kalamazoo system (Maureen Williams, VSE030, at Kayser Bondor, Merthyr Tydfil). In Silhouette's, Cardiff, the female factory-floor workers 'had little books and they'd write down what work they'd done and then they'd give me the books. All the books had to be made up to make their wages', says wages clerk Mary Brice (VSE048). Office workers, as staff on monthly salaries, could not earn bonuses or do piecework, and this meant that the assembly workers often earned more than they did, even though, as Betty Thomas (VSE046, Louis Edwards, Maesteg) says, 'I got the education!' Yet, they were also well aware that they had more freedom in the office environment and did not have to bear the pressure of the inexorable production line on the factory floor.

Pay Slips

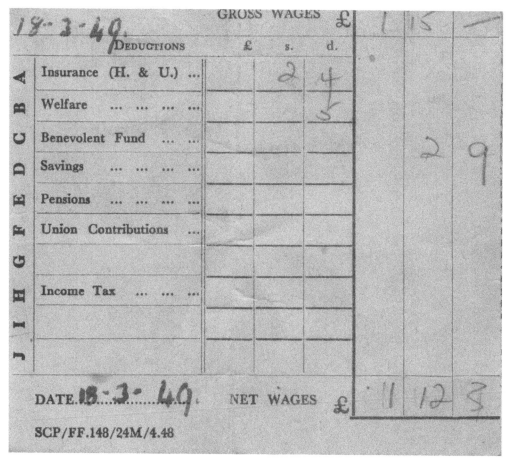

Irene Hughes (VSE024)'s first payslip from Kayser Bondor, Merthyr Tydfil, on 18 March 1949. (*Archif Menywod Cymru* Women's Archive of Wales)

J. R. FREEMAN AND SON LTD.

PAY SLIP   #344
                 FISH. F.D.

DATE	HOURS		EARNINGS						GROSS PAY
	BASIC	OVERTIME	BASIC PAY	OVERTIME PAY	PRODUCTION BONUS	HOLIDAY PAY			
DEC 3'65	40. 0.		9. 1. 3		2. 0. 0				11. 1. 3

GROSS PAY BROUGHT DOWN	DEDUCTIONS								NET PAY
	GRADUATED PENSION	INCOME TAX (REFUND BLACK)	SAVINGS CLUB	NATIONAL INSURANCE	BUS FARES	COMPANY PENSION	NATIONAL SAVINGS		
11. 1. 3	1.10	1. 1. 0CR	15. 0	11. 5					8.12. 0

JRF. 139.

PAY ADVICE		
WEEK OR MONTH NO.	DATE	39 2/1/76
DETAILS		
		21 30
A		6 90
B		
C		
D		
E		
GROSS PAY		28 20
GROSS PAY TO DATE		1125 80
TAX FREE PAY		510 90
TAXABLE PAY TO DATE		614 90
TAX DUE TO DATE		215 25
TAX REFUND		
TAX		5 35
GRADUATED CONTRIBUTION		
NAT. INS		1 55
1		
2		
3		
4		
5		
6		
TOTAL DEDUCTIONS		6 90
NET PAY		
F		
G		
TOTAL AMOUNT PAYABLE		21 30
EMPLOYER'S CONTRIBUTIONS	NAT. INS.	2 40
	H	
	J	
YOUR PAY IS MADE UP AS SHOWN ABOVE		Mrs. S. M. SMITH

Above: Frances Francis (VSE023)'s payslip from J. R. Freeman Cigars, Cardiff, on 3 December 1965. (*Archif Menywod Cymru* Women's Archive of Wales)

Left: Shirley Smith (VSE034)'s payslip from Burry, Son & Co., Treforest, on 2 January 1976. (*Archif Menywod Cymru* Women's Archive of Wales)

Factory Workers Together

Susie Jones (VN016) with fellow
workers at Cooke's Explosives,
Penrhyndeudraeth, in the 1940s.
(Susie Jones Collection)

Laboratory girls from Felinfach
Creamery, Ceredigion, enjoy a
lunch break in the 1950s. Meiryl
James (VSW053) is front right.
(Meiryl James Collection)

Tea break at Horrock's sewing factory, Cardiff, in the mid-1950s. (Maisie Taylor (VSE011) Collection)

A break in the Dunlop canteen for employees at Lastex Yarns, Hirwaun, in the early 1950s. Pat Howells (née Pendry, VSE082), sitting second left, is distinguished as the supervisor by the blue collar of her overall. (Pat Howells Collection)

The canteen at J. R. Freeman Cigars, Cardiff, in 1961. (J. R. Freeman photographs@Gallaher Ltd)

Margaret Gerrish (VSE080), far left, with coworkers at Corah Garment factory, Aberbargoed, Caerphilly, *c.* 1950. (Margaret Gerrish Collection)

Workers at H. G. Stone, Pontypool, in 1955. Jim Davies (VSE017) recalls that the girl at the front left was referred to as the 'flit' girl, who appeared regularly with a puffer – a spray insecticide to refresh the factory floor. (Dr Jim Davies Collection)

Camaraderie among laboratory girls at Llangefni Creamery, Anglesey, in the 1950s. Mair Griffiths (VN033) is centre back. Notice the acid burns in their laboratory coats. (Mair Griffiths Collection)

Marge Evans (VSE052) and coworkers at Sobell's, Hirwaun, in the 1950s. (Marge Evans Collection)

Camaraderie on a night out in Brecon for Ligne 10–12 assemblers at Anglo-Celtic Watch Company ('Tick Tock') in 1956. Meriel Leyden (VSW015) is centre front. (Meriel Leyden Collection)

Left: Mair Richards (VSW057), left, and Glesni Davies outside Pont Llanio Creamery, Ceredigion, in the early 1960s, with milk lorry and churns in the background. (Mair Richards Collection)

Below: Detonator workers at Cooke's Explosives, Penrhyndeudraeth, in the 1960s. (Susie Jones (VN016) Collection)

Above: Eira John (VSW048) and friends from the packing department at Mettoy's, Fforestfach, Swansea, link arms in companionship; 1960s. (Eira John Collection)

Right: Gwlithyn Rowlands (VN013) and fellow workers at Laura Ashley's in the late 1960s. (Gwlithyn Rowlands Collection)

Left: Three workers outside Lewis & Tylor (Tylore) (Gripoly Mills), Cardiff, in the 1970s. Jill Williams (VSE051) is in the middle. (Jill Williams Collection)

Below: Vicky Perfect (VN028)) and colleagues from Courtauld's, Flint, feature in a copy of a 1970s picture as a wall-drawing by graffiti artist, 'Random', in 2010. (Graffiti Artist 'Random')

Factory Products

Beti Davies (VN009) and Marion Davies
(VN010), left and centre, in their £5 coats,
made from wool produced at Glyn Ceiriog
Woollen Factory, Denbighshire, where they
worked, *c.* 1950. Beti says, '*they kept five
shillings a week from our wages until we'd
paid the £5*'. (Beti Davies Collection)

Mo Lewis (VN002) in a tea-towel dress from
Laura Ashley's, Carno, in the 1960s. 'The
first dress ever was ... two tea towels sewn
together. They just shaped the neck, shaped
the arms and we just sewed it together.'
(Mo Lewis Collection)

Left: A display of 'Glamor' buttons made by Welsh Products Ltd, Dowlais, at the Ryedale Folk Museum, Hutton le Hole, Yorkshire. (Ceinwen Statter and Dave Jones)

Below: A compact from James Kaylor Compacts, Caernarfon, *c.* 1960. The verse 'Take heed and see ye nothing do in vain. No minute gone comes ever back again' is painted around its border. (*Archif Menywod Cymru* Women's Archive of Wales)

Humpty Dumpty collection box made at High
Speed Plastics Ltd, Bangor, 1970. (Geoff Charles
Collection, National Library of Wales)

Factory Magazines and Advertisements

The front cover of the union magazine,
The Garment Worker, November 1952 issue.
(The GMB Union)

The back cover of the union magazine,
The Garment Worker, November 1952 issue.
(The GMB Union)

George Webb Shoe Factory Bulletin, Bridgend,
1957. (VSE037 Collection)

A Kayser Bondor advertisement, 1960s.
(oldmerthyr.com)

Advertisement for 'assemblers' to build washing machines at Hotpoint, Llandudno Junction, 1974. (Conwy Archives)

An advertisement for Berlei Bras; the photograph was taken at Big Pit colliery, Blaenafon, in the mid-1970s. The contrast between the old and new industries, the black and the white, is striking and symbolic. (Big Pit National Coal Museum)

Chapter 4

Health and Safety

Most of the women interviewed laughed when questioned about health and safety regulations in the period covered by the project, and dismissed the topic as irrelevant to their factory experiences. '*We were tough nuts – country bumpkins*', declares Nan Morse (VSW017). Yet, 75 per cent of the speakers questioned had stories to relate of accidents and mishaps, of poor working conditions and dangerous practices. The following reads as a catalogue of factory woes.

Factories were notoriously noisy places. Beryl Buchanan (VN055) recalls the horrendous din of fifty machines going 'zzz' all day at Mona Products, Menai Bridge, while the 'Bang! Bang! Bang! Bang!' at Alupack, Bridgend, throbbed in Alison Rees (VSW042)'s head even when lying in bed at night. It was the constancy rather than the level of the noise at Pembroke Woollen Factory that affected Mandy Jones (VSW067). Many factories, with their concrete floors and glass roofs, were also incredibly cold in winter and extremely hot in summer. No heating would be provided in many cases and the operators describe working in their coats, gloves and scarves, and adding whisky to their tea to warm themselves (Margaret Young, VSW065, Corgi's, Ammanford; Christie-Tyler worker, Bridgend, VSE038). Tools and equipment could also be difficult to handle under such circumstances, and Sally Cybulski (VSW008) froze because she had to go back and forth fetching lenses from huge freezers at the Optical, Kidwelly. May Lewis (VSW002) was so cold at Morris Motors, Llanelli, in the 1940s that she resorted to wearing her boyfriend's trousers to work, though she had never worn trousers before! In summer, temperatures could soar to 90°F (32°C) – '*you were cooked alive,*' says Dilys Wyn Jones (VN006) of the Corset Factory, Caernarfon. Although they tried to improve conditions by painting the glass roofs green, or by whitewashing them or hosing them down, working conditions remained uncomfortable. The management acknowledged the discomfort caused and free squash was distributed at Mettoy's, Fforestfach (Joan Gibbon, VSW060), while Sheila Hughes (VSE009) remembers being given salt tablets, too, to help restore her body fluids at British Nylon Spinners, Pontypool. At AB Metals, Abercynon, Sylvia Reardon (VSE006) was concerned about the office space. She read the Health and Safety Act, which noted that each person was entitled to 45 square feet each. She approached management and, when the workers arrived next day, they did indeed each have the required space because the ceiling of the office had been removed! Disagreement over working conditions often led to disputes, as will be seen in Chapter 5.

The machines on the factory floor, especially when workers disregarded instructions or when machines were faulty, could be dangerous, and numerous accidents are recorded. Dilys Pritchard (VN053) describes how a fellow worker at Austin Taylor's, Bethesda, ignored her advice, given as the leading hand on the assembly, to tie her long hair back. It was caught in the machine and it took Dilys half an hour to extract it rather than cut it all off. Dilys also witnessed the top of a finger flying past her and landing on the factory floor. She rescued it and its owner, who had not listened to instructions to put the safety cover on his circular saw, was reunited with his finger top. Similar stories pepper the interviews. Sandra Cox (VSE049) lost the tip of a finger at Freeman's Cigars, Cardiff, as did Mair Richards (VSE025) in the cutting machine at Kayser Bondor, Merthyr Tydfil. Her friend wrapped the bloody nail with the flesh underneath and ran around the factory shouting, 'Look, Mair has cut her finger off! And people were fainting!' The tops of one presser's fingers were left in the presser at Polikoff's, Rhondda, when she failed to pull her hand out quickly enough – 'it was quite horrific at the time,' says Anita Jeffery (VSE043). Yvonne Bradley's (VSW063) experience of having her arm dragged down by a conveyor belt at Morris Motors, Llanelli, epitomises the haphazard nature of health and safety regulations at this time. She passed out and a stretcher was fetched to carry her out, but

> when they went to pick it up, the stretcher part was still on the floor. They only had the handles, it had rotted. That's how safety conscious they were – they hadn't even checked the equipment ... So the two boys who were picking me up ... were arguing about ... who was going to have my legs, and who was going to have my tits ... By the time I came round they were still arguing. It was so funny.

Workers (Enid Davies and Bronwen Williams, VSW018) at the dirty, ramshackle but happy Deva Dog Ware factory in the late 1960s recall how the health and safety inspector was regaled with '*a little drink*' in the factory office, and never actually toured the factory itself.

By far the most common accident occurred on the powerful industrial sewing machines as machinists sewed through fingers and nails. Myfi Powell Jones (VN012) voices the universal adage among rag-trade workers – that one was not a fully qualified machinist unless you had experienced such an accident. In most cases, the machinists just pulled the needle out, dressed the wound and got on with their work (VN046, Eira Richards, Sewing Factory, Denbigh). At other times a mechanic or supervisor would perform the operation, before sending the patient to see the factory nurse. Sometimes, however, a piece of the needle broke off in the finger; Gwlithyn Rowlands (VN013) experienced this at Laura Ashley's, Carno, and she had to go for an X-ray at Aberystwyth hospital. Speakers also describe mechanics dismantling machines and taking both them and the patient to hospital for treatment (Anon. VSW013, Slimma's, Cardigan). Meirion Campden (VSW020), who worked at Glanarad Shirts, Newcastle Emlyn, asked whether she could keep the needle as a souvenir – but no, it had to be entered in the accident file. Most of the speakers blamed themselves for such accidents; one was called 'a clumsy cow' (VSW013), another called herself 'dopey' (Phyllis Powell, VSE016).

Although this was light industry, many of the workers worked with dangerous substances. At the top of this list must come the explosives produced at Cooke's Explosives, Penrhyndeudraeth. Two of the interviewees recall the horrific explosion of 28 August 1957, when three detonator girls and one boy were killed. Marion Roberts (VN011) who was young herself at the time, was traumatised by the incident, because she had just been talking to one of the girls involved before the fatal accident. Blodwen Owen's (VN026) testimony is even more poignant, because she lost her sister, Elizabeth, in this explosion. The inquest into the explosion declared that there was 'no negligence on the part of the company', and her family received no compensation as her sister had no dependents. The company only paid for the funeral, though there was nothing left to bury. Blodwen worked for this company for twenty-six years but had to retire early because of nitroglycerine poisoning. She also notes that the gelignite produced at Cooke's made workers faint and led to dependence upon it. During summer holidays from the factory, some suffered withdrawal symptoms and excruciating headaches and, in spite of the extreme danger of such an act, these workers would carry small pieces of gelignite in their pockets to sniff occasionally, so as to relieve themselves of their headaches (Iorwerth Davies, VN030; Beryl Jones, VN027 and VN026).

In other factories, employees worked with a range of noxious and even poisonous substances. Ann Owen (VSE005) was exposed to blue asbestos, flux and lead as a jewellery solderer at Attwood & Sawyer's, Porthcawl, and Doreen Bridges (VSE054) describes how she cut white asbestos for ironing boards at Golmet's, Pontllanfraith. 'You could see it in the air ... (you) didn't realise it was bad for you,' she declares. At Graesser's Salicylates, Sandycroft, Marilyn Hankin (VN048) had to wear a mask and gloves to handle the dangerous chemicals that went into aspirin and other tablets, but, she says, 'You got on with it, and if you don't like it, well, the door's there.' Other interviewees, Mavis Coxe and Sheila Edwards (VN049), describe a fatal accident at this plant, due to the volatility of phenol. The batteries made at Chloride Power, Pont-henri, contained cadmium, another chemical now listed as a hazardous substance. Although Gloria Brain (VSW005) was well aware of its toxicity, she wouldn't have dreamt of claiming compensation in case she lost her job. The use of glue and thinners introduced another hazard into the workplace. As soon as they opened a new tin of Bostik at Addis, Fforestfach, 'you'd be as high as a kite', according to Margaret Morris (VSW043); 'we'd all be laughing', says Mary Macdonald Davies (VN005), because of the glue at Llanberis Rubber Factory. Workers at Avon Inflatables, Dafen, were given frequent breaks and advised against glue sniffing (Beryl Jones, VSW050). Laboratory workers, by the very nature of their work, came into contact with all kinds of acids. Rosie James (VN032), Llangefni Creamery, and Evana Lloyd (VSW054), Felinfach, Cardiganshire, describe the effects of sulphuric acid on their skin, their eyes and their clothes. '*I remember coming home once,*' says Evana, '*I didn't have any tights left. They had completely disintegrated.*' At Kenfig Carbide, Betty Metcalfe (VSE056) worked with concentrated sulphuric, hydrochloric, hydrofluoric (which could eat through the porcelain dishes used) and nitric acid:

Your hands – you looked as if you were a heavy smoker. On the day I got married my fingers were almost yellow! I was ashamed to put my hand out to put the ring on.

However, immediate action could alleviate some of the problems, as Audrey Gray (VSE050), who worked in the laboratory at British Nylon Spinners, Pontypool, recalls:

> One of the flasks which we put on the hot plate, for some reason it just bubbled up and shot the solution and caught me on the face. But because we had a first aider down the corridor it was dealt with immediately. They took me to the hospital and I had no reaction at all, treated straight away.

Other environmental hazards were not so obviously dangerous to the health of the interviewees. The dust at the Flax Factory, Milford Haven, 'was like a fog', according to Poppy Griffiths (VSW039), while at the Corset factory, Caernarfon, the floor had to be watered regularly to keep the dust from the lime used in the making of the corsets down to a tolerable level (Dilys Wyn Jones, VN006). Pat Howells (VSE082) describes how the French chalk used at Lastex Yarns, Hirwaun, had to be blown off their overalls with a blowing machine five minutes before the end of every working day: 'it gave the workers bad chests', she says. Many suffered from skin problems too. The acid in the lollies at the Sweets' Factory, St Asaph, used to get under the workers' nails and build up into painful boils, recollects Joyce Edwards (VN045), while at Smith's Crisps, Fforestfach, the sharp edges of the crisps cut the hands and made them sore. Patricia Lewis (VSW028) had to buy her own gloves to protect her hands from the glass wool used at British Leyland, Llanelli, though the workers were provided with a leather apron and spats for welding. She also describes how a hot metal welding rod went into her chest and caused a deep burn; indeed, solder burns on hands, knees and legs were fairly common occurrences (Anita Roberts, VN036). Several of the speakers also comment on how they smelled even when they left work, due to the different substances used in the manufacturing processes. Workers at the Castle Works, Courtauld's, smelled of burnt toast, according to Vicky Perfect (VN028), and those at Fairweather Works, Pont-henri, of thinners (Nanette Lloyd, VSW004). Meanwhile, the smell of tobacco lingered on the clothes of workers at Freeman's Cigars, Cardiff (Ann Davies, VSE032).

As has been noted, some of the factories tried to address the issue of health and safety in various ways during the timespan of the period under scrutiny. Some provided protective clothing or footwear. When Nanette Lloyd (VSW004) was thrown across the factory floor because of an electrical fault at Fairweather Works, Pont-henri, the manager told her to go and get herself 'a pair of rubber wellies'. At the Corona factory, Porth, the workers had to wear clogs because of all the broken glass. They wore these out in their dinner hour too – 'they could hear us coming a mile away – everybody knew we were from Corona because of the clogs,' recalls Maureen Jones (VSE003). If a minor accident occurred, the usual response was that it was their own fault; they weren't concentrating properly or had been negligent. Nesta Davies (VN025), who worked at Johnson's Fabrics, Wrexham, sums up the general attitude: 'you didn't think about it [compensation] in them days.' Some, however, did claim for more significant injuries. Betty Probert (VSE053) remembers receiving £60 for an injury at Hoover's, Merthyr Tydfil, but was 'told not to apply ever again', while Margaret Duggan (VSE078) received £180 at Freeman's Cigars, which was a 'lot of money' in 1970, for jamming her hand in a machine. The loss of a finger secured compensation of £1,000 for a worker at James Kaylor Compacts, Caernarfon, according to Megan Owen (VN022). Today, most of the interviewees complain of suffering from such ailments as: tinnitus and acoustic neuroma due to levels of factory noise; carpal tunnel syndrome due to doing repetitive work; or osteoarthritis and deteriorating discs due to doing heavy demanding jobs for long periods. However, very few have sought or received any compensation for their long-term disorders, primarily because of the time lapse since the time they worked in a factory and their dislike of 'making a fuss'.

While the statutory regulations regarding health and safety were woefully inadequate until the introduction of the major Health and Safety at Work Act of 1974 and, while the evidence recorded above confirms this, many of the medium-sized and larger factories did employ full-time qualified nurses to deal with minor and major injuries from day to day. According to Pat R. D. (VN051), the surgery at de Havilland, Broughton, was like a mini health centre, with three sisters on duty and an ambulance ready for emergencies. Two industrial nurses, Lynfa Maeer (VSE065) and Enid Thomas (VSW066), were interviewed and, although they worked from the late seventies onwards, which is slightly outside the timeframe of this study, their testimony bears witness to the dangers and often atrocious conditions in many factories even in this late period. The interviewees appreciated this service greatly, and they note that the nurses also ministered to period pains (Yvonne Morris, VSE008), headaches due to the noise and to hangovers (Beryl Jones, VSW050), and helped with problems of the heart (Mary Brice, VSE048). Towards the end of the sixties and early seventies, too, there was increasing emphasis upon training first aiders on the factory floor and Christine J. Jones (VN043) says that at Pilkington's, St Asaph, she was required to follow courses such as those recommended by ROSPA.

Health and Safety

Detonators group at Cooke's Explosives, Penrhyndeudraeth, before a fatal accident in 1938 when three girls and two boys were killed. Susie Jones (VN016) is far left in the second row. A similar accident occurred, killing four, in 1957. (Susie Jones Collection)

Above and below: Workers at James Kaylor Compacts, Caernarfon, *c.* 1960. Notice the warning signs 'SAFETY FIRST!' and 'KEEP ALL YOUR ATTENTION ON THE MACHINE'. Yet Megan Owen (VN022) recalls a fellow worker losing a finger in a machine in the factory. (Dafydd Llewelyn Collection)

Right: Sonia Gould (VSE002) in her Denex sewing
factory overalls, with a bandaged finger, 1958.
(Sonia Gould Collection)

Below: The nurses' surgery at J. R. Freeman Cigars,
Cardiff, 1970s. (J. R. Freeman photographs@
Gallaher Ltd)

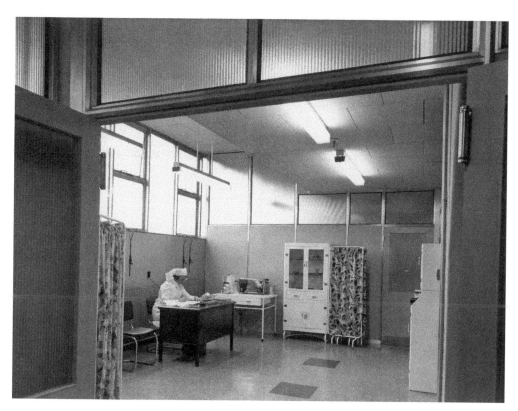

Factory Buildings

The iconic art deco building, Smith's Crisps, Fforestfach, Swansea; it was built in 1948. Currently INCA Creative occupies the building. (*Archif Menywod Cymru* Women's Archive of Wales)

Hoover's, Merthyr Tydfil, opened in 1948. (*Archif Menywod Cymru* Women's Archive of Wales)

The arch into Welsh Hills Works (Corona factory), Porth, Rhondda. (*Archif Menywod Cymru* Women's Archive of Wales)

James Kaylor Compacts factory, Turkey Shore, Caernarfon, 1960s. (Dafydd Llewelyn Collection)

Left: Ferodo's, Caernarfon, 1967.
Carol (VNo21) made stair treads
and brake linings in this car factory.
(Geoff Charles Collection, National
Library of Wales)

Below: The dining room at British
Nylon Spinners, Pontypool, 1950s.
(Pontypool Museum)

Chapter 5

Disputes and Strikes

Regarding the many aspects of factory life examined in this book, the views of the female workers, although they may have had vastly different working experiences, do concur on the whole. On the matter of disputes, strikes and most notably trade unionism, on the other hand, they often display different attitudes and influences. Well over half the interviewees showed little interest in the subject – it seems to have been a peripheral issue in their working lives. They weren't sure whether they had been members or not and, if they had been, they had difficulty naming the union. One worker, Meriel Leyden (VSW015) at 'Tick Tock', Ystradgynlais, admitted that she wasn't always sure why she was on strike, although she firmly believed that unions were important. Student Jenny Kendrick (VSE020) found the apathy difficult to fathom and noted that the women she met were downtrodden. Of course, in some factories, especially those of family-run companies such as Corgi's, Ammanford, and Laura Ashley's in mid Wales (VSW065, VN040 – workers remember Bernard Ashley's vehement 'No, no, no, no, no' to a request for a union), unionism was strictly prohibited, though some workers found ways around this ruling. Many interviewees commented that such intervention was not necessary because relationships with bosses and managers were always cordial. 'They were absolutely brilliant, they looked after their staff,' declares Doreen Lawson (VSE021) of the owners of Avana Bakery, Cardiff, in the early 1960s. Such sentiments are echoed by Mrs P. P. (VSE036) of her 'wonderful employers' at Hunt & Partners, Bridgend, and by Pat R. D. (VN051) of de Havilland, Broughton, which was, she maintains, 'a listening factory'. When workers at Gripoly Mills, Cardiff, discussed setting up a union, they came to the conclusion that 'we won't bother, for the simple reason being everybody's getting on so well; we sort out our own problems' (Jill Williams, VSE051). Yet, perhaps it should also be remembered that many of the workers interviewed were very young at the time, or were part-time married women who did not prioritise unionism or strike action. Their main aim was to earn their crust – 'you had to pay the union fees, which you didn't want taken out of your wages,' was Patricia White (VSW058)'s view. The fact that trade union meetings were often convened after factory hours in the evening was also inconvenient for this cohort (Mair Richards, VSE025).

However, this is only half the story. In reality, while women were poorly organised during the 1940s and 1950s, the situation did change dramatically during the following decades, as women became trade union members at an unprecedented rate. It was a struggle for some aspiring trade unionists. At Morris Motors, Llanelli, according to Gwen

Evans (VSW014), *'there was the most incredible fuss when unionism was first introduced'*
in the forties and she had to pay her dues of a 'groat' (4d) every three weeks, on the sly.
This evidence is corroborated by Yvonne Bradley (VSW063), who says that, even as late
as 1967, she had to go to the toilets to pay her subs. Two former workers (VSE035) at
Sidroy Mills, Barry, describe their efforts to introduce the Tailors' and Garment Workers'
Union into the factory in the 1940s and 1950s. After some initial obstacles, they found
themselves chair and secretary of the local branch and attending male-dominated TUC
meetings:

> During early meetings, we sat there speechless, until at one meeting in approximately
> 1953, we made our Maiden Speech. At that time, the Underwear Trade in Great
> Britain was experiencing a slump due to Competition from Japan ... My sister was
> the first to speak on the threat that cheap imports posed to our own industries which
> could lead to redundancies. I then spoke in support ... When we had finished our
> speeches, delegates clapped and shouted 'Here, Here' [*sic*] in the customary language
> of agreement.

These two workers are not typical of the majority, perhaps, but there are others
who established trade union branches and became adept at expressing their views
too. A worker (VSE038) at Christie-Tyler, Bridgend, decided to start a trade union
in that factory to address unfair treatment regarding overtime. Several interviewees
voiced their concerns that staff, as opposed to factory-floor workers, had no unions to
represent them. This was why Sylvia Reardon (VSE006) had to negotiate secretly in a
public house called The Junction and collect fees in secret among staff colleagues at AB
Metals, Abercynon, in order to establish a branch of the Clerical and Administrative
Workers' Union.

Undertaking union work could be a challenging venture. Vicky Perfect (VN028) was
only nineteen and the youngest representative when she was chosen to attend an All
Women Trade Union Conference in London in the late sixties. Meanwhile Edith Williams
(VSE062), 'a union girl' with the Hosiery Workers' Union, also travelled to London to
argue the case in Westminster for keeping the Kayser Bondor factory, Merthyr Tydfil,
open in the early seventies. This was to no avail, unfortunately, as it closed in 1977. Such
experiences widened their horizons. Being a union representative, says Anita Jeffery
(VSE043), 'gave me a boost really ... you were the chopsy one ... the one who spoke
for the girls'. In her case, however, she was offered a supervisor's job in order to curb
her tongue, an offer that she promptly turned down. Others were not as self-confident.
Margaret Tegwen John (VSE014) had to be persuaded by her colleagues that she would be
the ideal person to represent female issues with management on their behalf, and Yvonne
Bradley (VSW063), in spite of her obvious loquaciousness as a speaker, felt that her
written skills were not up to the mark. As trade unions became stronger in the sixties and
seventies, in many cases they also became closed shops. When Grace Beaman (VSW035)
began working in *c.* 1968 at Unit Superheaters, Swansea, she was told unequivocally,
'If you don't join you'll be out of here before your feet can touch the ground, good girl –
no question about it. I'll go and get the manager and march you off now,' while at Berlei
Bras, Pontardawe, in the 1970s, 'If you didn't pay your union the others could go on

strike and say "We're not working with a (non)-union person"' (Anon. VSW030). Sonia Gould (VSE002) confirms that at Denex, Tredegar, workers had 'no choice' but to be members of the National Tailors' and Garment Workers' Union and she remembers the union meetings in the factory canteen. This didn't worry her because 'they stood by you in any dispute.'

The issues that caused disagreement have been touched upon already. Shift work could be contentious and attitudes varied from factory to factory. At Courtauld's, Flint (Vicky Perfect, VN028), and Chloride's, Pont-henri (Gloria Brain, VSW005), female workers were not allowed to work nights and, at British Nylon Spinners, Pontypool, they were considered 'too delicate' to do so (Sheila Hughes, VSE009). On the other hand, Gwen Evans (VSW014) did work on the night shift at Morris Motors, Llanelli, for a while, but disliked the experience. Shop steward Mair Matthewson (VSW046) also worked all three shifts for about ten years at Metal Box, Neath:

And then these do-gooders, if you want, came – it was too much they said – they didn't want women doing shift work, right? They weren't supposed to do the night shift and the women had to just do two shifts then – mornings and afternoons, but the men did three shifts, see ... I wasn't willing at all. ... When they stopped women working night shift the men had to learn how to do the open tops too. This happened in 1963.

She also describes how they drew all the women off the feeding machines and replaced them with men. '*Why? Because someone had probably moaned that things were too difficult for the women, after years of doing it. So the women had to teach the men how to do the work*!' she declares disdainfully. In these cases, the unions do not seem to have intervened on the women's behalf and every factory interpreted the regulations as they wished. The unions did, however, fight for shorter working weeks on behalf of their members. Meriel Leyden (VSW015) attributes the reduction from a 44- to a 40-hour week at 'Tick Tock', Ystradgynlais, to union intervention, while at Freeman's Cigars, Cardiff, workers of the Tobacco Workers' Union walked out of the factory in the seventies at 1.30 p.m. one Friday, in order to force the factory to adopt this new finishing time for its workers. Eventually the owners conceded but, as Margaret Duggan (VSE078) says, this is the only dispute she recalls during the thirty-two years she worked there.

Working conditions could bring workers out on strike, as at Polikoff's, Rhondda, when a plague of rats and black pats over-ran the bedding department in the 1950s (Anita Jeffery, VSE043). At Denex, Tredegar, the call 'All out! All out!' rang through the factory if the workers were not allowed to listen to the radio, or if it was too hot or too cold, according to the thermometer on the factory wall. 'We ran the factory really, we did,' maintains Sonia Gould (VSE002), and Yvonne Bradley (VSW063) echoes these comments based upon her experiences in Morris Motors in the early 1970s. 'We had the power. The workers had the say in them days ... We ruled the work.' Even before unionism was introduced, because there were so many factories in some areas desperate for female operatives, the women did have considerable negotiating power. When Marge Evans (VSE052) and her fellow operators at Sobell's, Hirwaun, were refused extra bonuses, they threatened that they would all go to work at Murphy's factory nearby. They knew they had the managers 'over the barrel' because they were the only operatives capable of working the big machines at Sobell's. On the other hand, when Cynthia Rix (VSW052) and seventeen other machinists went on an unofficial strike at Windsmoor's, Fforest-fach, in the early 1950s because they were refused a pay rise, they were sacked and their immediate response was to join the workforce at Mettoy's nearby. Likewise, Sylvia Howell (VSW062) walked out of John Stanton's, Llanelli, because her work had been regraded from that of a top machinist to an ordinary machinist in order to pay her less; she left the factory early one Friday afternoon and started at Salter's on the following Monday. Some of the interviewees describe attitudes towards female blacklegs during strikes in this period. In view of its turbulent striking tradition, arising from the great Penrhyn Quarry strike of 1900–03, it is hardly surprising that workers at Austin Taylor's, Bethesda, viewed strike breakers with particular animosity:

> *The lives of those who had gone into work wasn't very pleasant when everyone else had returned; they didn't talk to them … those who had stayed out and had lost money refused to work with those who had gone in during the strike* (Sandra Owen, VN054).

Eirlys Lewis (VSW061) experienced such treatment herself after she and one male colleague refused to support a strike at Vandervell Products, Cardiff. She felt that the managers had done their utmost to solve the problem of the intense heat of the exceptionally hot summer of 1976. Part of one wall was removed as well as the windows from the roof, and industrial fans were brought in. Eirlys argued that they couldn't just '*switch off the sun!*' As a result, she was called all sorts of names, put '*into Coventry for four months*' and some never spoke to her again. She concludes that the women were worse than the men. Eirlys was exceptional in other ways as well. She refused to pay the Labour Party levy incorporated in her trade union fees because, she says, politically, she is a Welsh Nationalist. However, when another interviewee (Mary Lynn Jones, VSW045) expressed the same view to her union representative at 'Tick Tock' in the sixties, she was ignored and told, rather patronisingly, that she was '*splitting hairs*'.

As might be expected, the main bone of contention was the unequal pay women received. The famous strike of the female upholstery stitchers at the Ford Dagenham plant in 1968 helped ensure the Equal Pay Act of 1970, which came into force in December 1975, stating that women and men should have equal pay for equal work. Yet, very few of the interviewees seem to have been aware of the significance and implications of this vital piece of legislation, although some were aware of the tactics employed by both managers and male union officials, such as demanding that they work night shifts or that they fulfil heavy manual tasks and refusing to help them, to try to discourage women from claiming equal pay. During her first years as a factory worker, says Sandra Owen (VN054), there was a difference between men and women's work but, once the Equal Pay Act was passed, she had to work in every department of the machine shop at Austin Taylor's, doing the same work as the men. However, it is

Marion Blanche Jones (VSE028), a shop steward in the 1970s at Hoover's, Merthyr Tydfil, who offers us the best account of the experiences of female factory workers regarding the issue of equal pay during this period:

> The Equal Pay Act came in in 1970. It came by law and we still weren't on equal pay. And the women said 'Well, you know they've had it in Ford's, what's wrong with Hoover's?' ... So we started the ball rolling, we got in touch with Ann Clwyd, and she helped us a lot, mind, advising us which way to go. We approached management and they said 'Well, we've got no qualms about the equality you know. It's been brought in and we're in agreement ...' But of course the [union] convenor stepped in then and said the women weren't doing the work the same as the men ... It started this animosity and the men decided to go on strike, although it was against the law. They were on strike for two days – they lost two days' pay over it. But they had to give in in the end ... It wasn't a very pleasant time at all ... not at all ...
>
> Once Equal Pay came in, they expected women to work on lines. I ended up on the tumble dryer line, horrible jobs, but the men used to come around and say, 'Well, you're having the same money, you're expected to do the same job.' So they'd throw anything at us.

When the BBC came to interview her about the men's strike, she explained, 'We're not fighting the firm, we're fighting the union, Hoover's union, and the shop stewards.' The women stuck together at Hoover's but, unfortunately, 'the bad feeling lasted years'.

Celebrations and Social Life

An official photograph of the staff of J. T. Morgan stores, Swansea, with the workers at Glanarad Shirt factory, Newcastle Emlyn, on their annual outing to Llandrindod Wells. It is dated 26 November 1948. J. T. Morgan and his brother, Johnny, owned both concerns. (Rita Davies (VSW020) Collection)

Staff of J. T. Morgan stores and Glanarad Shirt factory workers on a joint outing to Cilgwyn Manor, Newcastle Emlyn, *c.* 1950. (Rita Davies (VSW020) Collection)

The Social and Sports Club Entertainments' Committee, Coldrator's, Llandudno Junction, *c.* 1950. Kathy Smith (VN023) is front right. (Kathy Smith Collection)

Christmas dinner at Cooke's Explosives, Penrhyndeudraeth, with the male workers serving the female workers, in the 1950s. (Susie Jones (VN016) Collection)

Horrock's workers at a dinner dance in the Connaught Rooms, Cardiff, 1955. Rita Spinola (VSE001) is on the right. (Rita Spinola Collection)

Workers and partners from Christie-Tyler (upholstery), Bridgend, at their sumptuous dinner in Cardiff City Hall in the 1950s. (Ernest Carver & Son)

Workers at George Webb Shoe factory, Bridgend, performing a cabaret act during a Christmas party in the 1950s. (VSE037 Collection)

Christmas dinner at Laura Ashley's, Carno, in the late 1960s. Bernard Ashley sits at the head of the further table and Laura Ashley sits third down on his left. (Gwlithyn Rowlands (VN013) Collection)

It's party time for the workers of Club Shirts, Neath Abbey, in this view from the 1970s. (Skewen Industrial Heritage Association)

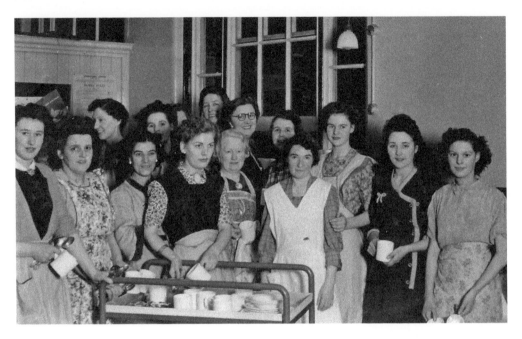

Preparing the children's Christmas tea party at Courtauld's, Flint, in the 1940s. Vera Jones (VN042) is fourth from left, front row. (Vera Jones Collection)

The annual Christmas party for the children of workers and staff at Anglo-Celtic Watch Company ('Tick Tock') in the 1950s. (Joyce Evans (VSW027) Collection)

Playing games at a Christmas party for the children of workers and staff at Ferodo's, Caernarfon, in the 1960s. (Carol (VN021) Collection)

INA Bearings' Christmas party for workers' children, held in Loughor Welfare Hall, in the mid-1970s. Beryl Evans (VSW025) stands on the right. (Bonfoto Cine Studio, Bryn, Llanelli)

Workers at Anglo-Celtic Watch Company ('Tick Tock'), with some of the toys they bought with fellow workers' donations for children at Killay Children's Home, Swansea. Christmas 1954. (Mair Williams (VSW068) Collection)

South Wales Switchgear's Christmas sleigh and company carol choir, which travelled the area, in a view from the early 1960s. (Hawker Siddeley Switchgear)

Chapter 6

A Collective Sub-Culture

The regimented nature and inexorable quality of factory work would suggest that female workers endured long, dull days in their working environment. Yet, this was not always the case, as the workers sought to alleviate boredom and monotony in many different ways. Anna Pollert, in her book *Girls, Wives, Factory Lives* (Macmillan, 1981, p. 21 and p. 130,) claims that such women developed a shop-floor culture where they could demonstrate

> dignity and defiance in the face of being tied to a machine and being 'put down' as 'just' women factory workers ... Women replaced the rule book ... into an informal code of resistance to being turned into machines, into boredom, to the humiliation of being ordered around ... they did this within very tight limits – without interfering with production – machines rattled on.

The testimony of many of the interviewees recorded for this project, although they were not always aware of the significance of their actions, confirms Pollert's assertions and reflects the fact that there was an unwritten collective factory-floor subculture, which enlivened life and made the work more tolerable. As student Michele Ryan (VSE070) says of her fellow workers at the Glass Works, Cardiff, 'They were bolshie, always having a laugh.'

This manifested itself in many ways, and this chapter will try to capture these memories and this atmosphere, while also remembering that all factories did not operate in the same way and that attitudes varied and changed over the years. Talking was strictly forbidden in some factories; the supervisor would be 'there like a shot' at B. S. Bacon toy factory, Llanrwst, says Nancy Denton (VN019), and 'down on you like a ton of bricks' at George Webb, Bridgend (Anon. VSE037). The owner, Johnny Morgan, used to smack the girls on their heads (with a pencil) and sometimes pinch them if they broke this rule at Glanarad Shirts, Newcastle Emlyn, according to Rita Davies and Meirion Campden (VSW020) – yet, '*You respected him*', they conclude. Some workers chose not to chitchat. Keeping your line going and not letting your fellow assemblers down meant total concentration, and those on piecework felt it interfered with their ability to earn more money. Even when there was no ban imposed, the noise often made talking very difficult, but not impossible. As supervisor Eileen Davies (VSW026) comments about the workers at Slimma's, Llandovery, '*Don't worry, they found a way of talking – these were women!*' One anonymous speaker (VSW030) at the Economics factory, Pontardawe notes that 'You could always tell factory girls because they were loud' and the butt of remarks such as 'Listen to that lot!' from their fellow office workers. The usual remedy, however, was to learn to lip read and many became expert at it (Patricia White VSW058, Lotery's, Newport). Sheila Hughes (VSE009), who worked at British Nylon Spinners, Pontypool, describes how she communicated effectively 'with girls at the other end of your lab, which must have been 100 feet long and about 40 feet wide ... by lip-reading'.

Opinions are divided among the interviewees about the amount of swearing on the factory floor, although the perceived view was that 'language' was part of this culture. At London Pride, Bridgend, contends Gwen Richardson (VSE018), they were very strict and workers could be sent home if caught. Workers at 'Tick Tock', Ystradgynlais (Catherine Evans VSW016), Lastex Yarns, Hirwaun (Pat Howells, VSE082), and Kayser Bondor, Merthyr Tydfil (Mair Richards, VSE025), paint a similar picture, but the last speaker contrasts this dramatically with her experience of a supervisor at AB Metals, Abercynon, whose 'every other word would be the F word'. Unsurprisingly, Mair stayed at ABs for only nine days! Others took a more philosophical stance. Sandra Brockley (VN050) did find the women at Courtauld's 'rough' initially, especially as she was not used to women swearing, but 'did join them after a while'; student Susan Leyshon (VSE063) soon got used to the colourful language at Revlon's, Maesteg, though she did not participate herself.

Listening to music and singing helped to alleviate the tedium greatly. Many factories allowed suitable music programmes to be relayed through a tannoy system. Interviewees fondly remember *Workers' Playtime, Housewives' Choice* and *Music While You Work* in the fifties and sixties. The latter was aimed specifically at making factory workers more productive. Ann Owen (VSE005) of Attwood & Sawyer's, Porthcawl, agrees: 'We worked better with music. You cut it off ... it slowed production down.' They list their favourite artists, too, among them Nat King Cole, Pat Boone, Dickie Valentine, and later Elvis, Cliff and Englebert Humperdinck. When Gene Pitney sang her favourite song, Enid Davies (VSW018) describes how she '*connected*' with her friend across the factory floor at Croydon Asbestos, Milford Haven. A worker at Denex, Tredegar, wrote 'I love Elvis' in chalk all the way down a trouser leg – she wasn't sacked, but she did have a warning (Sonia Gould, VSE002). Workers sang along with records and the radio, and there can be little doubt that this was a valuable education in popular culture for many of the young girls, who might not have heard these artists at home. Others note that Welsh hymns and songs were popular at 'Tick Tock' (Meriel Leyden, VSW015) and Deva Dog Ware, Gwynfe (Nan Morse, VSW017). Melva Jones (VSW012), who obviously had a good singing voice, was always being urged to entertain her fellow workers with a song at J. R. Hargreaves Pop, Llanelli. The interviewees also revel in the fact that some of Wales's most renowned singers worked in manufacturing at this time – among them were Dame Shirley Bassey (Curran's, Cardiff)[7]; Irene Spetti, with the stage name of Lorne Lesley, 'who was a scream' and would have them all singing at Horrock's, Cardiff, claims Rita Spinola (VSE001); and Iris Williams, at Planet Gloves, Llantrisant, who 'always wanted to sing something operatic', says Gwen Richardson (VSE018), but adds, 'we used to drown her out with a pop song.'

Due to the demands of the non-stop production of the assembly line, the operators often had to raise a hand to ask permission to go to the toilet or wait for a particular 'reliever' to come to free them, whether they wanted to go or not. In many cases, they were also timed. At Courtauld's the supervisor would be knocking on the door, says Vicky Perfect (VN028), and Luana Dee (VSE015) recalls someone shouting 'Have you finished?' at her at Berlei Bras, Merthyr Tydfil. The toilets often became centres of mild rebellion against such authoritarian attitudes; as Lynfa Maeer (VSE065) says, 'it was quite a social thing in the toilets' at Polikoff's, Rhondda. She recalls having her hair put up into a beehive there and several interviewees describe washing their hair on Friday afternoons to go out in the evening. Others (Sonia Gould VSE002, Denex, Tredegar; Patricia Murray, VSW019, Alan Paine, Ammanford and Cynthia Rix, VSW052, Mettoy's, Fforestfach) had their ears pierced in the factory toilets. Cynthia sat on the toilet to have the piercing done, another girl minding the door, while another put the needle through a flame. When she went home, her mother 'battered' her. It was in the toilets at James Kaylor Compacts, Caernarfon, that Megan Owen (VN022) learned to do the quickstep and the waltz. The girls could smoke here, too, and this 'fag break' was tolerated by some supervisors if they weren't allowed to smoke at their work stations, as they were in heavy industrial works, such as at Davies Steel, Pembroke Dock (Mandy Jones, VSW067). At Addis, Fforestfach, however, smoking was a 'massive issue' says Yvonne Morris (VSE008), because the supervisors followed the smokers into the toilets to flush them out. Glenys Hughes (VN041) was really annoyed at the timewasting cigarette breaks at Laura Ashley's: 'I should finish two years early for all the cigarette breaks I didn't take,' she says.

Perks played an important role in making factory life and pay more tolerable, though not all factories were generous to their workers, as Yvonne Stevens (VN003) laments of B. S. Bacon toy factory, Llanrwst. She would have liked to have been able to afford a toy fort or a skipping rope for her younger siblings, but they were beyond her means. On the other hand, at Mettoy's, workers could buy a big bag of cars for only £5, says Margaret Hayes (VSW033). Clothing factories could afford to sell their seconds cheaply. Thus, Caroline Aylward (VSE055) managed to buy two Marks & Spencer 'not quite perfect' duster coats at Louis Edwards, Maesteg, in the 1950s and another worker at the same factory (VSE068) could remember paying only £1 for a dress. Larger items were available, too. Betty Probert (VSE053) got a washing machine, cleaners and a tumble drier when she worked at Hoover's, and she continues, 'I can go down the factory now if I want to, to get anything.' She also notes that the workers were allowed to eat as many chocolates as they wanted when she worked at OP Chocolates. British Leyland workers were even luckier, as they could have a discount of about £2,000 off a new car worth around £5,500 (Patricia Lewis, VSW028) but, as Yvonne Bradley (VSW063) so succinctly says, 'The trouble is you couldn't afford it ... If I couldn't afford bus fare, I'm damned sure I couldn't afford a car.' Two other examples of factory perks that epitomise the period are the 200 packet of cigarettes workers received at Freeman's Cigars (Ann Davies, VSE032) and Beryl Buchanan (VN055)'s comment that the only freebie they got at Mona Products, Menai Bridge, were reject knickers to use as sanitary towels.

Pilfering was a sackable offence and, at some factories, such as High Speed Plastics, Llandygái, (Sandra Owen, VN054) and Freeman's Cigars (Sandra Cox, VSE049), they employed security staff to randomly search workers leaving the factory. Yet, despite Freeman's vigilance, according to Frances Francis (VSE023):

> A lot of the women came from the valleys. Oh and the valley girls always wore curlers … you'd never see a valley girl without hair curlers. And apparently they used to hide the cigars in the hair curlers, and some of them had been caught because we used to be searched.

Some workers at Slimma's, Cardigan would roll up completed trousers, put an elastic band around them and carry them out past security guards in their sandwich tins (Anon. VSW013), while at Smith's sewing factory, Rhumney:

> They'd start on the line by here, and by the time they got down to the bottom of the line, there was two jackets missing … they were taking parts so when they'd go home they'd sew them up … and make a jacket up. (Sonia Gould, VSE002)

Zips (Averil Berrell, VSW034, Lightning Zips, Waunarlwydd) and end-of-line cotton reels (Sylvia Howell, VSW062, John Stanton's, Llanelli) were commandeered as required too.

One fascinating aspect of factory subculture was the innocent and not-so-innocent mischief making that took place. Operators would send sweets or notes to each other down the line. At Mettoy's, one note urged the women to 'Smile if you had it last night!' (Margaret Morris VSW043), while in other factories notes would be posted to customers in the packing process, such as in tablecloths at Fairweather Works, Pont-henri (Nanette Lloyd (VSW004). Betty Probert (VSE053) recalls how they placed a message, 'If married, pass me by. If single, please reply' in chocolate boxes at OP Chocolates and how her sister received a reply from a Dutch merchant seaman, with whom she later spent a week's holiday at his home. New young factory girls were teased through harmless initiation rites. The interviewees recount how they would be sent to the stores to fetch all manner of impossible things, such as sky hooks at Davies Steel, Pembroke Dock (Mandy Jones, VSW067); a long stand at Austin Taylor's, Bethesda (Sandra Owen VN054); elbow grease at Ferranti's, Bangor (Enid Jones VN052); a left-handed screwdriver at 'Tick Tock' (Meriel Leyden, VSW015); or a bucket of steam and striped cotton at Windsmoor's, Fforestfach (Cynthia Rix VSW052). Cynthia notes that the experience 'upset me something terrible … but, I got over it.' This could be construed as low-level bullying, as student Christine Chapman (VSE067) suggests, but on the whole it was accepted as the norm and a deviation from the tedium of everyday life.

The women had to contend, too, with a more insidious culture that was common in the period studied, where there was a challenging mix of male and female workers. Wolf-whistling (Christine J. Jones VN043), sexual comments such as those Carol (VN021) endured at Ferodo's, where the man responsible received 'a slap on the wrist' for his temerity, and pinching bottoms at Polikoff's, according to Maureen Howard Boiarde (VSE031), were common occurrences. Yet attitudes varied greatly towards such incidents. Maureen maintains that they were to be expected – 'actually … if you didn't have your bottom pinched, you'd go home and you'd cry! You'd think there was something wrong with you!', while Isabel Thomas (VSE040), who worked at Mansel Tinplate in the forties, felt that the teasing broadened the minds of the innocent girl workers. Other women relate tales of unpleasant personal harassment they experienced as office workers: one was at Polikoff's, Rhondda, where a 'randy old man' used to put his arms around her and pester her (VSE042), and another at Penn Elastic, Fforestfach, when it was suggested to the interviewee that there was a particular route to gain promotion to work in their London office – a route that she was not prepared to take (VSW032). On the other hand, several speakers describe being looked after by the older men in the factories. They were very fatherly, says Evana Lloyd (VSW054) of her fellow male-workers at Felinfach Creamery, while at de Havilland, Pat R. D. (VN051) felt like 'I had a million dads looking after me'.

The men interviewed for this project throw a different light upon such incidents, and several of them claim that their female colleagues were just as capable of sexual harassment when they, as young boys, ventured into the all-female space on the factory floor. Dafydd Llewelyn (VN007), who worked at James Kaylor Compacts recalls the women shouting '*Do you want a thrill?*' at him when he was just fifteen, and this is corroborated by Bill Moses (VSE058), who worked as a salesman in many factories. At Gossard's, Pontllanfraith, 'I'll have your trousers down! I'm waiting!' were some of the women's verbal threats, he says. Nanette Lloyd (VSW004) disapproved of intimidating comments such as '*You watch out, I'll strip you and tie you up!*' that were aimed at new male workers when she worked at Fairweather Works, Pont-henri. Male interviewee VSE081, who worked on maintenance at Sidroy's, Barry, during the sixties, actually experienced the reality of such a threat. He and several other young male employees were 'de-bagged' (i.e., stripped) by the female employees. Another speaker, Keith Battrick (VSW064), describes a similar incident, where both male and female workers were involved in what appears to be an initiation rite. It was 1966 and he was an aspiring apprentice at London Transformers, Bridgend. It was, he says,

a very traumatic experience for someone who is just sixteen, just left school, not particularly worldly wise. ... I was taken by a couple of men, and frogmarched to one part of the factory where there were women and lots of other people gathered around. And this particular ceremony that they did there was they would actually strip you off ... I was stripped off and then – women were taking part as well – they would smear you with grease around your private parts, and sawdust, and just leave you ... I can remember being just left with obviously everyone making rude remarks and enjoying themselves.

Although Maureen Howard Boiarde (VSE031) declares of her colleagues at Polikoff's, 'We were the ballsiest, gutsiest females that God put on the earth.' Only two of the cohort of female interviewees recall such incidents and it is interesting to note that it is a temporary student worker, Michele Ryan (VSE070), who provides one of these descriptions from her observations at the Glass Factory, Cardiff:

I remember one young bloke starting on the line ... and there was one big central track though the factory floor, and they said, 'Oh you've got to be introduced to everybody, so we'll line up and you can just go through the line'. He was mortified, he was really shy. ...

'It's alright, we're all going to introduce ourselves.' Of course, as he went through, the women, every single one of them, either pinched his bum or made some comment about his body, grabbed his balls! I mean there was a real initiation. That was a really good way of … making sure that nobody messed with them.

The other description comes from interviewee VSW063 who organised initiation parades for new boys at Morris Motors, Llanelli. When they came up to her, she would tell them to bend over, and she would give them a little 'tickle', and say 'You're all right boy, you can go.' No wonder they left her with red faces, even though, she says 'I never touched them, it was only a little tease.' This ceremony, a vital example of collective factory subculture, seems to have helped these women to stamp their authority on the alien world of the factory floor.

More Celebrations and Social Activities

The line-up for the Miss Dunlop contest in 1957. Pat Howells (née Pendry) (VSE082) stands fourth from the left. (Pat Howells Collection)

Above: Anita Jeffery (VSE043) winning the Miss Polikoff competition in the late 1950s. (Anita Jeffery Collection)

Right: Sandra Harding (later Cox) (VSE049), who won Miss Freeman (Cigars) in 1974. (Sandra Cox Collection)

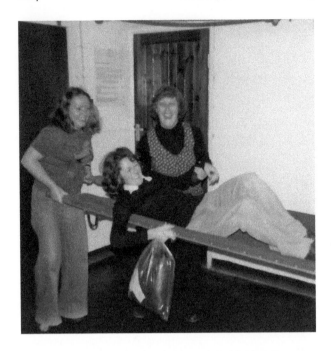

Manhandling Gwlithyn Rowlands (VN013) by carrying her on a stretcher before dipping her in the pond to celebrate her birthday at Laura Ashley's, Carno, in the 1970s. (Gwlithyn Rowlands Collection)

Maisie Taylor (VSE011) with the wedding dress made for her by her fellow workers at Horrock's sewing factory, Cardiff, in 1957. (Maisie Taylor Collection)

Right: Fellow workers at Felinfach Creamery, Ceredigion, form a guard of honour in clean new overalls, holding their laboratory test tubes, at a wedding in Ciliau Aeron Church in the early 1960s. Meiryl James (VSW053) is first left. Note the churn in the foreground. (Meiryl James Collection)

Below: J. R. Freeman's Cigars' 'It's a Knock-out' team in the 1970s. (J. R. Freeman photographs@Gallaher Ltd)

Above: Laura Ashley's football team in the 1970s; Gwlithyn Rowlands (VN013), Olive Jones (VN039) and Glenys Hughes (VN041) were members. (Gwlithyn Rowlands Collection)

Left: Gwynedd Lingard (VSE061), who worked in BSA Tools, Cardiff, and represented Great Britain as a gymnast in the Olympic Games in Helsinki, 1952, and Rome, 1960. (Gwynedd Lingard Collection)

Right: Pat Perks, who worked at J. R. Freeman's Cigars, Cardiff, was an Olympic gymnast who competed in the Rome Games, 1960. (Catrin Edwards Collection)

Below: Metal Box, Neath's jazz band, 1950s. Mair Matthewson (VSW046) is fourth from the right, front row. (Mair Matthewson Collection)

Above: A fancy dress competition at Laura Ashley's, Carno, in the 1970s. (Olive Jones (VN039) Collection)

Left: Nanette Lloyd (VSW004) and friend dressed in fancy dress made from remnants from John Patterson (Fairweather Works) tablecloth factory, Pont-henri, Carmarthenshire, in the mid-1950s. The factory decorated a float for the village carnival. (Nanette Lloyd Collection)

Blodwen Owen (VN026) with fellow workers at Cooke's Explosives, Penrhyndeudraeth, on her retirement in 1970. (Blodwen Owen Collection)

Meriel Leyden (VSW015) receiving a gold watch made at Anglo-Celtic Watch Company ('Tick Tock') from the managing director, Mr Bolt, in 1980, to mark twenty-five years of service at the factory. (Meriel Leyden Collection)

Above: A retirement gift for Margaret Duggan (VSE078) on leaving J. R. Freeman's Cigars, Cardiff, 2002. It states, 'It is intended for use as a placebo type treatment for those long boring days when you have nothing to do.' (*Archif Menywod Cymru*, Women's Archive of Wales)

Left: The official opening of INA Bearings, Bynea, Llanelli, with Jim Griffiths, who was MP for Llanelli and Secretary of State for Wales, the manager and Beryl Evans (VSW025), who was on inspection on the assembly line, in 1966. (Bonfoto Cine Studio, Bryn, Llanelli)

Workers waiting outside for Queen Elizabeth's official visit to Hoover's, Merthyr Tydfil, 1973. Marion Jones (VSE028) is far right, second row. (John Yates)

Vera Jones (VN042) among male colleagues on Courtauld's Works' Committee, Flint, in the 1940s. (Vera Jones Collection)

Left: Cooke's Explosives, Penrhyndeudraeth Works' Committee, in the 1970s; Iorwerth Davies (VN030) is second from right, second row from back. (Iorwerth Davies Collection)

Below: Jeanette Groves (VSE077), standing on the left, and colleagues from Western Shirt Company, Cardiff, on their annual paid trip to Bourton-on-the-Water in the late 1940s. They were given a packet of cigarettes each for the journey! (Jeanette Groves Collection)

Workers from Kayser Bondor, Merthyr Tydfil, on a day out to Weston-super-Mare in the late 1950s. Irene Hughes (VSE024) is on the left. (Irene Hughes Collection)

Deva Dog Ware workers on a trip to London in 1967. Nan Morse (VSW017), second right, is wearing a pair of white boots bought on the trip and a dress she made for herself for the outing. However, she says that London was a dirty place and that the dress was filthy when she got home. (Nan Morse Collection)

Workers from INA Bearings, Bynea, on a trip to Blackpool in 1975. Beryl Evans (VSW025) is fourth from right in the back. (Beryl Evans Collection)

A good night out for the workers at Windsmoor sewing factory, Swansea, in the 1960s. Patricia Ridd (VSW041) sits fourth from the left. (Patricia Ridd Collection)

A night out for Cardwell factory workers, Lampeter, in the mid-1970s. (Augusta Davies (VSW011) Collection)

Chapter 7

Celebrations and Social Activities

While most of the interviewees emphasise that their main aim in working in a factory was to earn money, they also almost invariably note that the camaraderie engendered there is their one abiding memory of the factory floor. Surprisingly, very few talk of female cliques or of bitchiness. Most of them enjoyed working with other women and *'making friends you could trust'*, to quote Dilys Pritchard (VN053). This camaraderie could begin with the bus journeys to work. Laura Ashley's provided their own transport (Mo Lewis, VN002) and workers (Margaret Tegwen John, VSE014; Yvonne Smith, VSE042) remember dozens (the number varies) of double-decker buses serving Polikoff's in the Rhondda. Many of the women could remember the exact times of buses to work in the morning, because this was of daily importance to them. Others, obviously, walked and filled the roads with their chatter on such trading estates as Treforest and Hirwaun. At the end of the day, however, it was a mad rush to leave – *'we went like animals going out of a zoo'*, says Meriel Leyden (VSW015) of the 'Tick Tock' workers.

One factor that helped Meriel to settle in well at 'Tick Tock', she says, was the amount of Welsh spoken there, which made her feel at home, and similar comments were made about Morris Motors (Glenda and Annie Lewis, VSW003), Slimma's (Helena Gregson, VSW009) and Felinfach (Meiryl James, VSW053) – all factories in south-west Wales. Indeed, it could be argued that these and other similar factories in south-west and north-west Wales helped to keep the Welsh language a vibrant community language in their areas by providing employment for Welsh-speaking women. On the other hand, Margaret Hayes (VSW033) implies that at Mettoy's, Fforestfach, Welsh speakers from Pontarddulais always spoke English on the factory floor and that 'they would be clouted' if they spoke in their native tongue. The suggestion that this was a managerial policy, however, is not borne out by any other testimony, but Eirlys Lewis (VSW061) does recall being challenged for 'speaking that foreign language' at Vandervell Products, Cardiff. Her reply was short but to the point: 'We are speaking our own language in our own country.' Another comment in a similar vein is that made by Madeline Sedgewick (VSE079), who was surprised that she managed to get a job at all at Curran's, Butetown, Cardiff, because she had heard that 'they were very racial ... they were colour prejudiced.'

Celebrations helped to raise spirits and alleviate the monotony of everyday work. Marriages were celebrated with particular ceremony, not unlike the initiation rites of

Chapter 6. Anita Roberts (VN036) describes her experiences at the Capacitor Factory, Wrexham, in the 1970s:

> They grabbed hold of me and they tied me with tape and they put me in this trailer, which had wheels and a handle on it and they covered me in flour and sprayed stuff all over me, like shaving foam, and I was absolutely covered, and they put me on the main road, on the pavement, and left me there ... It was quite funny really, everybody was looking at me and beeping their horns as they went past.

Likewise, when Olive Jones (VN039) got married in 1971, the workers at Laura Ashley's, Carno, put her on a trolley and tipped her into the river, in the presence of Laura Ashley herself! Anita Jeffery (VSE043) describes the rite at Polikoff's, Rhondda:

> The girls on your conveyor belt or on your section would backcomb your hair, put all sugar soap in it ... and all the chalk that they used for marking out the materials ... and put you in a truck and roll you down into the men's department, and leave you there.

At Morris Motors both the aspiring bride and bridegroom would be put on a pallet and covered with flour, eggs, curry powder, jam and condoms (Yvonne Bradley, VSW063). The workers at 'Tick Tock' devised a slightly different rite. Moira Morris (VSW001) describes how they all dreaded being dragged into the toilets and dressed in head-dress and veil, with confetti stuffed down their clothes, and made to walk up and down the factory floor with a mop and bucket, while the women sang bridal songs such as 'Here Comes the Bride' and 'I'm Getting Married in the Morning'. A similar custom involving the dressing of a veil and tying bits of wool in the hair prevailed at Corgi's, Ammanford, too, according to Margaret Young (VSW065), but could also be seen when one of the female workers celebrated a birthday. They would be tied to a chair, their faces blackened with coal, bits of their hair tied with wool, then placed on a trolley and left in the car park, where the men of Teddington's factory next door could see and mock them during their break times. In contrast, the practice of forming a guard of honour at a colleague's wedding, which was customary among Felinfach Creamery workers (Meiryl James, VSW053), seems very genteel.

With such a mix of male and female workers in the same working environment, it is hardly surprising that the factories often served as marriage bureaux. Examples could be multiplied, as in the case of Kitty Jones (VSW056), who could count at least four other friends who had found husbands at Hancock's Brewery, Swansea. Some of the larger factories printed their own newsletters or magazines, and these marriages were faithfully recorded in them. Among the newsletters mentioned by interviewees are *Smoke Signals* (Freeman's Cigars, Margaret Duggan, VSE078) and the *Rayoneer* (Courtauld's, Vera Jones, VN042). They also recorded births and deaths and the kaleidoscope of other social activities associated with factory life. One popular annual feature would be the crowning of the latest factory queen. We hear of Miss Pilkington (Christine Jane Jones, VN043), Slimma's Queen (Helena Gregson, VSW009), and the Miss Manikin contest at Freeman's Cigars, which 'always caused trouble every year', says Tryphena Jones (VSE074). Supervisor Pat Howells (VSE082) went around Lastex

Yarns, Hirwaun, urging girls to compete and ensuring they had suitable bathing costumes to wear. 'There was always a good turnout,' she says. While Pat never won, another interviewee, Anita Jeffery (VSE043), had fond memories of coming runner up and then of winning the Miss Polikoff's competition in the late 1950s – but she was chosen off the dance floor in her evening gown. Associated with these events might be the fashion parades of lingerie organised by the Berlei Bras (Luana Dee, VSE015) and Kayser Bondor (Anne Amblin, VSE022) factories to demonstrate their wares to prospective clients and staff, with suitable music and choreography. These were held either in the factory canteens, on the shop floor itself or in such aspirational shops as David Morgan's, Cardiff.

   Events such as the beauty competitions would often be run by the Works' Councils or the Sports and Social Clubs found in many factories. Contributions to the social funds would be compulsory, and the money was taken out of the workers' pay packet every week, says Patricia Murray (VSW019) of Alan Paine's, Ammanford. The larger factories had their own social clubhouses, such as at British Nylon Spinners, Pontypool, which Sheila Hughes (VSE009) describes with enthusiasm. It had one of the largest ballrooms in south Wales, with a sprung floor, a grand piano and a huge stage. There was also a rifle range, a judo club and a couple of bars. Sheila saw

> *The Wild One* before it was issued, with Marlon Brando. And we went to see John Ogden, pianist, (Alfredo) Campoli, the violinist, and … Léon Goossens, the oboe player. And they had people like Ted Heath, Joe Loss, Ken Cooper … all the big bands there.

The range of activities on offer here was staggering, as Audrey Gray (VSE050) notes:

> Every sport was catered for: tennis, bowls, shooting … skittles and sailing club, table tennis, billiards; there wasn't anything you couldn't do really. I did basket work, I did sailing club, I played occasional tennis, they had teams of football.

Sports, both for men and women, were very popular and much more diverse than might be expected. The women played hockey on 'a pitch like a ploughed field' at Kayser Bondor (Irene Hughes, VSE024); baseball at Hancock's Brewery (Kitty Jones, VSW056); badminton at the Royal Ordnance Factory, Pembrey, where Maureen Jones (VSW021) was taught by her bosses and had time off to attend tournaments; and netball at Dunlop's and Lastex Yarns, Hirwaun, where the factory engineers made the posts for the team (Pat Howells, VSE082). Pat also ran an archery class and was a member of a tug-of-war team. Dolgarrog Aluminium sported tennis courts and a swimming pool (Mair Williams, VN018) and from its Social Club at Hafod, Mettoy's, ran a darts team, which competed every Friday evening (Grace Beaman, VSW035). The Hoover's sports days were 'wonderful' and contestants would come from their other factories in Cambuslang, Scotland, and Perivale, London, to compete (Betty Probert, VSE053). Football was popular at the Laura Ashley factories and, according to Gwlithyn Rowlands (VN013), the company paid for their green kit, emblazoned with the Welsh dragon. On one occasion it sponsored a trip to London to play on a field near Wormwood Scrubs, says goalie Glenys Hughes (VN041). EMI and Polikoff's in the Rhondda organised football teams in 1965 to raise money for the families of the thirty-five victims of the Cambrian colliery disaster. EMI won but, in the Polikoff's Club after the game, 'there were ructions', says EMI fullback, Patricia Howard (VSE029), 'Polikoff's pushing and shoving – jealous because we had won.' A mark of the late 1960s was the popularity of the 'It's a Knock-out' competitions organised by several factories such as Laura Ashley's, Freeman's and Morris Motors. Local carnivals were also avidly supported. Workers from Switchgear, Pontllanfraith, prepared an annual Christmas float (Doreen Bridges, VSE054), and Sonia Gould (VSE002) of Denex organised the local carnival and chose the apposite theme, 'The Rag Trade', for their factory float.

One of the more remarkable sporting stories, however, involves two Cardiff gymnasts, Pat Perks and Gwynedd Lingard (VSE061). Pat Perks worked at Freeman's and, when she was chosen for the British Olympic team to compete in the Rome Olympics of 1960, the workers collected on the factory floor to buy her a uniform and equipment (Ann Davies, VSE032). Gwynedd Lingard competed in two Olympic Games – Helsinki, 1952, and Rome, 1960. Before the Helsinki Olympics she worked for Boots, the chemist, but they would not allow her time off to go to the games. Therefore, she left and joined BSA Tools, Cardiff, who gave her time off with pay. This was incredibly important for someone who came from Grangetown, 'very much a working class district', and who had to pay all her own sporting expenses. She says,

> When I was in my local environment I was a star, because I was going to be an Olympian. But when I was in work, I was really one of the girls.

Other social activities were organised, too, such as the factory choir at Ferodo's, Caernarfon (Carol, VN021); drama competitions between Dunlop's factories (Pat Howells, VSE082); a jazz band, dressed in white with red coats and marching kazoos in Metal Box, Neath (Mair Matthewson, VSW046). Pauline Moss (VSE019) was introduced to the opera in Cardiff. This range of activities shows how influential the factories were in both work and play, and how much they could contribute to the societies in which they were embedded. Not everyone chose to participate, of course, and many of the smaller factories had no such facilities or activities to offer their workers, including Cardwell's, Machynlleth (Phyllis Eldrige and Olga Thomas, VSW010); the Belt Factory, Ynysmeudwy (Anon. VSW029); and Glyn Ceiriog Woollen Mill. (Beti Davies, VN009)

Annual outings were also organised either by the workers themselves or by the management. Factories in north Wales flocked to see the Blackpool lights, and Porthcawl and Barry were favourite destinations in south Wales. Margaret Chislett (VSE012) has fond memories of transport on the double-decker buses and rides on donkeys. Others organised trips with sister factories, such as the special Swansea Hancock's Brewery train, which would pick up employees from Cardiff and Newport on its way to London (Kitty Jones, VSW056). These outings widened the workers' horizons and taught them many social skills. The main celebrations, however, were reserved for Christmas. Often the factory itself and individual machines would be decorated with balloons and trimmings, until health and safety regulations intervened. Many of the interviewees describe how work was suspended around midday on Christmas Eve, and drinks, smuggled onto the factory floor, would be consumed, so that 'by one o'clock we was all sloshed', laughs Sonia Gould (Denex, VSE002). Marjorie Collins (VSE026) admits that she got tipsy on sherry at Lines/Tri-ang Toys, Merthyr Tydfil, and fell into a tub of toy wheels! At 'Tick Tock', the workers drank in the toilets and, when the foremen tried to stop them, they were dragged into the wash room and had their hair washed. These are further examples of how the workers grasped at opportunities to subvert the customary factory discipline; the supervisors and charge-hands knew it was wiser not to interfere. The interviewees also describe the marvellous dinners prepared in the factory canteens to celebrate Christmas; in Freeman's the managers dressed up as waiters to serve their workers – thus turning the established order upside down once more (Margaret Duggan, VSE078).

Christmas evening parties, dinners and dances are described with compelling charm by the interviewees, and were obviously the highlight of their social calendar. Greta Davies (VN004) says that it was as if the workers at Aykroyd's, Bala, had been released (from jail) after all their hard work, and the parties in Neuadd y Cyfnod were fantastic. Horrock's, Cardiff, held its dinner dance in 1955 at the Connaught Rooms, and Rita Spinola (VSE001) remembers having her first evening gown and special shoes for the occasion. City Hall was one venue for AB Metals' fabulous Christmas dinners – another 'dress up and elbow gloves and silver shoes' affair, says Sylvia Reardon (VSE006). These were large factories, but most of the other factories held Christmas dos too, paid for by the staff themselves, or sometimes by the bosses. At the traditional Welsh-language celebrations for Felinfach Creamery workers, Meiryl James (VSW053) wrote home-spun comic verses and songs about her fellow workers. All kinds of raffles and hampers (Eirwen Jones, VSW055, Pont Llanio Creamery, Ceredigion) were organised for the festival, and interviewees who worked at Slimma's (Helena Gregson, VSW009) received a frozen turkey and a bottle of champagne each, if the company had reason to celebrate. The season of goodwill manifested itself in other ways, too. Many social clubs or works' committees organised Christmas parties with presents from Father Christmas for employees' children, and collections were made and toys delivered to children's homes. Interviewee Kathy Smith (VN023), who was in the personnel department at Hotpoint, Llandudno Junction, describes how she would send food parcels to sick employees and how 'The Good Companion' committee would deliver logs made from factory pallets. Yvonne Bradley's (VSW063) observation, 'We used to have some fun, we did; we worked hard and we played hard', encapsulates the spirit of the social life of the factories.

Last Strands

The last day at the Alan Paine factory, Ammanford, 1998. (Patricia Murray (VSW019) Collection)

Marge Evans (VSE052), third from right, with her ex-colleagues in a reunion of Sobell's workers in the 2000s. (Marge Evans Collection)

Chapter 8

Other Strands and Endings

The 1948 Holiday with Pay Act, which gave many factory workers the right to a fortnight's holiday with pay every year, was warmly welcomed and appreciated by the interviewees. This holiday period often coincided with the miners' fortnight – the last week in July and the first in August, when production would cease while essential maintenance work was carried out. On the day before the annual closure at Slimma's, Lampeter (Augusta Davies, VSW011), drinks flowed on the factory floor as they did in many factories on Christmas Eve. In the austerity and rationing of the forties and early fifties, few interviewees went on holiday at all, except to stay with family. However, gradually they began to venture to the holiday resorts at Blackpool, Llandudno and Porthcawl, and to the popular Butlin's camps in the late fifties and early sixties, and then further afield to the continent in the late sixties and seventies. Some were a little more adventurous. Kathleen Matthias (VSE069) notes that she had to work for a year before she was entitled to a holiday but, at sixteen years old in 1956, she ventured to Blackpool, followed by trips in 1958 to Majorca and in 1960 to Italy with friends from Kayser Bondor, Merthyr Tydfil.

One significant feature of women's factory careers was its discontinuity. In general terms, women would enter factory work in great numbers in their teens, stay into their mid-twenties, withdraw to bring up a family until their thirties, and then return in strength to work part-time and later full-time until retirement age. The older workers note that not many newly married women continued to work (Sali Williams, VN056), although no speaker actually experienced the marriage bar. Rather, the great majority of the interviewees say that the crucial leaving time was when they were six-and-a-half-months pregnant. '*That's it, full stop*', says Beryl Buchanan (VN055); their jobs would not be kept open for them and there was no such thing as maternity leave or pay. The better employers treated their employees with consideration during their pregnancies, e.g., moving them to less stressful work, as at Hunt & Partners, Bridgend, where Mrs P. P. (VSE036) went from making large packing boxes to wedding-cake boxes. Supervisor Pat Howells (VSE082) moved her pregnant girls to a special bench in Lastex Yarns, Hirwaun, but, as a result, of course, everyone on the shop floor also knew who was pregnant. Mary Evans (VN008) took pregnancy in her stride and continued working on the big presses at James Kaylor Compacts, Caernarfon, throughout. Indeed '*the press shook my body all day, for 8 hours,*' she says, '*and I didn't have any labour pains ... when I had my baby. It had done the work for me.*' Being forced to leave work caused great distress for several speakers – 'I was very, very upset. I broke my heart actually,' comments a worker at Morris Motors (Anon. VSW007). This sentiment is echoed by Evana Lloyd (VSW054), who took a year to settle after leaving because she missed the company of her family of friends at Felinfach Creamery. These women lost their independent incomes and missed the camaraderie of the factory. Home work or out-work was sometimes provided for homebound mothers by some employers, such as Laura Ashley's (Mo Lewis, VN002) and Deva Dog Ware, Gwynfe (Enid Davies, VSW018), though often at lower rates of pay for long hours. Long years of service were rewarded with the customary gifts of clocks and golden watches (Anon., VSE038, Christie-Tyler, Bridgend), paid for by the company, and on retirement a special leaving do would usually be organised. As 'the mother of the factory' at Hotpoint, Personnel Officer Kathy Smith (VN023) arranged the factory's retirement parties and ensured that the Director was present to say goodbye.

The variety of terms used naturally by the factory workers reflect the different tasks they performed, and also the hierarchy within the workplace. Most of the women recorded were factory-floor operatives (riveters, solderers, etc.), assemblers or machinists/top machinists, but others were cutters or pressers, or on inspection, ensuring the quality of the product before it went to be packed. If Greta Davies (VN004) found needle damage on garments when on inspection at Aykroyd's Pyjamas, Bala, she would mark the spot with a red sticker and return it to the machinist, for which Greta would be '*damned to the clouds … and told to put the sticker on her mouth*'. Large factories like Mettoy's employed fifty women in the warehouse 'putting up' toys in crates, as Eira John (VSW048) explains. There would be charge-hands and supervisors, often called 'white coats', in charge of groups of the women. They would have to know the production process well to be able to assist a worker in difficulty. Mair Williams (VN018) notes how awkward it could be to supervise her previous coworkers at Dolgarrog Aluminium, but she told them, 'You can guarantee I'll never ask you to do anything that I wouldn't do myself,' and thus won their respect. Patricia Murray (VSW019), on the other hand, was advised by her boss to 'be friendly … [but] to keep your distance', at Alan Paine, Ammanford. According to Pat Howells (VSE082), she had to wait until she was twenty-one before she could be promoted from supervisor to forewoman at Lastex Yarns, Hirwaun. However, most of the interviewees showed little ambition to join the managerial staff and, even if they had such aspirations, in most cases, as Yvonne Smith (VSE042) says, 'Women didn't climb up the ladder because they didn't get the chance.'

The division between the factory floor and the office staff could be marked, too. On the factory floor, 'we were all in the same boat – nobody posh', claims Vanda MacMillan (VN020, B. S. Bacon, Llanrwst), but some speakers describe superior attitudes because office workers were considered more educated (Luana Dee, VSE015, TSB, Merthyr Tydfil; Dr Jim Davies, VSE017, Weston's Biscuits, Llantarnam). Surprisingly, Ann Harrison (VSW040) asked to be moved from office work to the factory floor, even though she had the relevant qualifications, at Mettoy's because 'we had to know our place', she says. This divisive attitude was not reflected among the office workers interviewed, though they emphasised the differing nature of the tasks they performed as clerks, or in the wages or invoicing departments. They were skilled at typing and shorthand, and they note with pride their use of the comptometer, an early calculator. 'I was always up for a challenge' says Betty Thomas (VSE046, Louis Edwards, Maesteg). Sylvia Reardon (VSE006) 'worked like a dog' in the invoicing department at AB Metals, Abercynon, and there was such a rapport there that 'you felt that you were a very important cog in that wheel'. Several of the interviewees (e.g., Audrey Gray, VSE050, and Sheila Hughes, VSE009, at British Nylon Spinners, Pontypool) worked in factory laboratories and mastered technical processes and terms. Other interviewees who held challenging jobs in male environments were Maureen Jones (VSW021) as a tracer in a male drawing office at ROF Pembrey and Pegi Lloyd Williams (VN015), who became a buyer, and managed men on the factory floor at Metcalfe's, Blaenau Ffestiniog. Peggy was interviewed on television because it was unusual for a woman to become a buyer for an engineering firm in this period.

In his 1983 article 'Women workers in Wales, 1968–82'[8], historian Gwyn A. Williams surmises that 'the evidence suggests that, if the rate of change-over (from male to female labour) proves to be that of the years since 1968, women will form a majority of the working class at work in Wales by the end of 1994.' This prophesy was not to be fulfilled, however; during the last decades of the twentieth century, due to globalisation, mechanisation, new technologies and a host of other complex issues, most of the factories mentioned in this book closed, were taken over, and/or moved their production overseas. To give a few examples: Castle Works, Courtauld's closed in 1977; Mettoy's closed in 1982; Louis Edwards closed in 1981; British Nylon Spinners was taken over by ICI but ceased production in 1988; and all manufacturing by Steinberg Alexon was moved overseas in 2003. Empty factories littered the landscape or were demolished; as Shirley Smith (VSE034) says of the Treforest Estate she once knew, 'now it's gone, it's all derelict there.' The perception that redundancy was somehow less traumatic for married part-time or full-time female workers is shown to be utterly misleading. It was just as depressing and demoralising for women to be out of paid work as men. Helena Gregson (VSW009) cried for a week when Slimma's, Lampeter, was closed in 2001, but did progress to open her own company based upon the sewing skills she had acquired at the factory.

In summing up their experiences in the factories, many speakers claim, even when they didn't really enjoy the work, that these were the happiest days of their lives: 'Factory time is the best part of your life' (Mary Farr, VSE010, H.G. Stone, Pontypool); it was 'really home from home' (Jill Williams, VSE051, Gripoly Mills). They also maintain that it was '*a good education*' (Nan Morse, VSW017), not only regarding work-related skills but also for character building. Maisie Taylor (VSE011), Horrock's, says that 'It was the making of me, it brought me out, made me what I am today', and Eirlys Lewis (VSW061) testifies of her time at Pullman's Flexolators, Ammanford: '*you learned to live with people, of all sexes and all types*'. 'To work in a factory', asserts VSW030 (Anon.) 'you have to stand up for yourself ... because if you don't, the other workers will crucify you', and this is echoed by Ann Harrison (VSW040), who says, 'you could never be shy after working in a factory full of women.' The temporary student workers interviewed confirm this view of factory life. Jacqueline Jenkins (VSE013), who worked at the Royal Sovereign Pencil Factory (Staedtler) in Pont-y-clun, maintains that 'it was an eye opener ... it gave me that bridge to working life', while former Welsh Assembly Member Christine Chapman (VSE067) asserts, 'I think that [it] was an early feminist lesson ... It has been quite influential on my growing up and my career ... that was the real world really.' Likewise, the temporary and permanent male employees interviewed appreciated the factory subculture; as Dr Jim Davies (VSE017), who worked in several factories in the eastern valleys in his holidays, says, 'I broadened my understanding of human life ... considerably, and that stood me in good stead for the rest of my life.'

Virginia Nicholson, in her study of women's lives during the Second World War, concluded that this had been 'a generation of brave, stoical, unselfish, practical and uncomplaining women whose values, along with their deeds, seem to be passing into history'.[9] This is a very apt description of the interviewees of this project; nevertheless, in this case, they have been given a voice to ensure that their stories and experiences do not pass unheeded into history. Their voices are clear and eloquent and they speak for themselves:

> [Factory work] gave them [the women] a sense of independence and also a sense of who they were as individuals. Because most of them were married with families everybody [outside the factory] related to them in that way – they were somebody's wife and somebody's mother, but while they were in the factories they were themselves, and ... they got back in touch with their identity and who they were.
>
> Yvonne Morris, VSE008

> I've done my bit for John Jones' country, I think.
>
> VSE045, a worker at Louis Edwards, Maesteg

> At my age I feel like I haven't really achieved anything big in my life, but well I suppose I did contribute ... to the country really. I was a manufacturer.
>
> Anita Jeffery, VSE043, Polikoff's and Christie-Tyler

> I would never knock Hotpoint because it gave me my life, gave me my house, gave me cars, holidays and the kids a good upbringing ... It's a shame really there's no industry like that round here for the youngsters today, 'cos they'll never get what we had.
>
> Margaret Evans, VN037

End Notes

Chapter One

1. Beddoe, Deirdre, *Out of the Shadows* (Cardiff: University of Wales Press, 2000); John, Angela, *Our Mothers' Land* (Cardiff: UWP, 1991); Masson, Ursula, *'For Women, for Wales and for Liberalism' Women in Liberal Politics 1880–1914* (Cardiff: UWP, 2010).
2. Williams, Mari A., *A Forgotten Army* (Cardiff: UWP, 2002), p. 222.
3. Percival, Geoffrey, *The Government's Industrial Estates in Wales 1936–1975*, PhD dissertation, 1978, held at Glamorgan Archives (DIEC/29), p. 48.
4. Ibid. Appendix xiv.

Chapter Two

5. Elliott, Arthur, *History of British Nylon Spinners* (Abertillery: Old Bakehouse Publications, 2009), p. 44.
6. www.traditionalmusic.co.uk/folk-song-lyrics/Duw_Its_Hard

Chapter Six

7. Williams, John L., *Miss Shirley Bassey* (London: Quercus, 2010), pp. 60–61.

Chapter Eight

8. Williams, Gwyn A., 'Women Workers in Wales, 1968–82' in *Welsh History Review* Vol. 11, No. 4 (December 1983), p. 547.
9. Nicholson, Virginia, *Millions Like Us Women's Lives during the Second World War* (London: Penguin, 2012), p. xviii.

Acknowledgements

Thanks are due to: the management group and members of Women's Archive of Wales *Archif Menywod Cymru* for their continuous support and enthusiasm for this project; Heritage Lottery Fund for its generous funding; the Ashley Family Foundation, Unite and Community Unions for their kind donations; Field Officers Catrin Edwards, Susan Roberts and Kate Sullivan, for their professionalism and commitment to the project; and to all the volunteers who helped with recording speakers in the field.

Our greatest thanks, however, must go to the interviewees themselves, who gave so unstintingly of their time and their reminiscences to make this project such a valuable and worthwhile experience. This book is dedicated to them, I hope they enjoy it and that it does them justice.

All the material collected for this project has been deposited in the National Screen and Sound Archive at the National Library of Wales, Aberystwyth. Transcriptions of most of the interviews conducted can be seen on our website www.factorywomensvoices.wales or www.lleisiaumenywodffatri.cymru.

Picture Credits

The author and publisher would like to thank the following people/organisations for permission to use copyright material in this book: the Geoff Charles Collection, National Library of Wales; Skewen Industrial Heritage Association; J. R. Freeman photographs@ Gallaher Ltd; Glamorgan Archives; David Garner; Paul Norton photographer, Glamorgan Archives; Pontypool Museum; Hawker Siddeley Switchgear; Corgi Hosiery; Richard Firstbrook, Llandeilo; Ernest Carver and Son; the GMB Union; oldmerthyr.com; Conwy Archives; Big Pit National Coal Museum; Bonfoto Cine Studios; John Yates; 'Random' graffiti artist; Ceinwen Statter and Dave Jones.

Thanks are also due to the following for use of photographs from their collections: Pontyberem Historical Society; Dafydd Llewelyn; Ann Swindale; Catrin Edwards; Caroline Aylward; Averil Berrell; Yvonne Bradley; Carol; Marjorie Collins; Sandra Cox; Ann Davies; Augusta Davies; Beti Davies; Iorwerth Davies; Jim Davies; Nesta Davies; Rita Davies; Beryl Evans; Joyce Evans; Marge Evans; Margaret Gerrish; Sonia Gould; Audrey Gray; Mair Griffiths; Poppy Griffiths; Jeanette Groves; Gaynor Hughes; Irene Hughes; Pat Howells; Meiryl James; Anita Jeffery; Eira John; Marion Jones; Olive Jones;

Rosie Jones; Susie Jones; Vera Jones; Eirlys Lewis; Mo Lewis; Meriel Leyden; Gwynedd Lingard; Nanette Lloyd; Mair Matthewson; Nan Morse; Patricia Murray; Blodwen Owen; Mrs P. P.; Vicky Perfect; Mair Richards; Patricia Ridd; Gwlithyn Rowlands; Violet Skillern; Kathy Smith; Rita Spinola; Maisie Taylor; Isabel Thomas; Patricia White; Jill Williams; Mair Williams; VSE037; *Archif Menywod Cymru* Women's Archive of Wales.

Every attempt has been made to seek permission for copyright material used in this book. However, if we have inadvertently used copyright material without permission/acknowledgement, we apologise and we will make the necessary correction at the first opportunity.